Geometry

Applying • Reasoning • Measuring

Chapter 7
Resource Book

The Resource Book contains the wide variety of blackline masters available for Chapter 7. The blacklines are organized by lesson. Included are support materials for the teacher as well as practice, activities, applications, and assessment resources.

McDougal Littell
A HOUGHTON MIFFLIN COMPANY
Evanston, Illinois • Boston • Dallas

Contributing Authors

The authors wish to thank the following individuals for their contributions to the Chapter 7 Resource Book.

Eric J. Amendola
Patrick M. Kelly
Edward H. Kuhar
Lynn Lafferty
Dr. Frank Marzano
Wayne Nirode
Dr. Charles Redmond
Paul Ruland

ISBN: 0-618-02070-5

123456789-VEI- 04 03 02 01 00

Contents

Contents

Contents

Descriptions of Resources

This Chapter Resource Book is organized by lessons within the chapter in order to make your planning easier. The following materials are provided:

Tips for New Teachers These teaching notes provide both new and experienced teachers with useful teaching tips for each lesson, including tips about common errors and inclusion.

Parent Guide for Student Success This guide helps parents contribute to student success by providing an overview of the chapter along with questions and activities for parents and students to work on together.

Prerequisite Skills Review Worked-out examples are provided to review the prerequisite skills highlighted on the Study Guide page at the beginning of the chapter. Additional practice is included with each worked-out example.

Strategies for Reading Mathematics The first page teaches reading strategies to be applied to the current chapter and to later chapters. The second page is a visual glossary of key vocabulary.

Lesson Plans and Lesson Plans for Block Scheduling This planning template helps teachers select the materials they will use to teach each lesson from among the variety of materials available for the lesson. The block-scheduling version provides additional information about pacing.

Warm-Up Exercises and Daily Homework Quiz The warm-ups cover prerequisite skills that help prepare students for a given lesson. The quiz assesses students on the content of the previous lesson. (Transparencies also available)

Activity Support Masters These blackline masters make it easier for students to record their work on selected activities in the Student Edition.

Alternative Lesson Openers An engaging alternative for starting each lesson is provided from among these four types: *Application, Activity, Geometry Software,* or *Visual Approach.* (Color transparencies also available)

Technology Activities with Keystrokes Keystrokes for Geometry software and calculators are provided for each Technology Activity in the Student Edition, along with alternative Technology Activities to begin selected lessons.

Practice A, B, and C These exercises offer additional practice for the material in each lesson, including application problems. There are three levels of practice for each lesson: A (basic), B (average), and C (advanced).

Contents

Reteaching with Additional Practice These two pages provide additional instruction, worked-out examples, and practice exercises covering the key concepts and vocabulary in each lesson.

Quick Catch-Up for Absent Students This handy form makes it easy for teachers to let students who have been absent know what to do for homework and which activities or examples were covered in class.

Cooperative Learning Activities These enrichment activities apply the math taught in the lesson in an interesting way that lends itself to group work.

Interdisciplinary Applications/Real-Life Applications Students apply the mathematics covered in each lesson to solve an interesting interdisciplinary or real-life problem.

Math and History Applications This worksheet expands upon the Math and History feature in the Student Edition.

Challenge: Skills and Applications Teachers can use these exercises to enrich or extend each lesson.

Quizzes The quizzes can be used to assess student progress on two or three lessons.

Chapter Review Games and Activities This worksheet offers fun practice at the end of the chapter and provides an alternative way to review the chapter content in preparation for the Chapter Test.

Chapter Tests A, B, and C These are tests that cover the most important skills taught in the chapter. There are three levels of test: A (basic), B (average), and C (advanced).

SAT/ACT Chapter Test This test also covers the most important skills taught in the chapter, but questions are in multiple-choice and quantitative-comparison format. (See *Alternative Assessment* for multi-step problems.)

Alternative Assessment with Rubrics and Math Journal A journal exercise has students write about the mathematics in the chapter. A multi-step problem has students apply a variety of skills from the chapter and explain their reasoning. Solutions and a 4-point rubric are included.

Project with Rubric The project allows students to delve more deeply into a problem that applies the mathematics of the chapter. Teacher's notes and a 4-point rubric are included.

Cumulative Review These practice pages help students maintain skills from the current chapter and preceding chapters.

LESSON 7.1

INCLUSION Ask students to identify some examples of the three basic transformations that occur in the classroom or that they might have experienced that day. Examples as simple as the opening and closing of the classroom door or window (rotation about a point), horizontal or vertical movement of student desks (translation), and looking in a mirror at home (reflection in a line) can help the understanding of all students, including those with limited English proficiency.

COMMON ERROR Students may tend to think that the three basic transformations identified in this lesson are the only isometries or rigid transformations. Be aware that glide reflections in Lesson 7.5 also belong to this group.

TEACHING TIP Students should understand that the order of the vertices in the name of the figure is important when using arrow notation to indicate one figure mapping onto another. This is mentioned on page 397. The correspondence of points in a mapping is just as important as it is in the identification of congruent triangles.

LESSON 7.2

TEACHING TIP Remind students that the paper folding activities on page 403 require accuracy and neat work because the folded paper will be used to investigate the measures of angles and segments.

COMMON ERROR Some students may think line symmetry is the only type of symmetry. Others may recognize that you could turn figures, like those in parts (b) and (c) of Example 4 on page 406, and the figure would map onto itself. Be aware that rotational symmetry, which is also called point symmetry or turn symmetry, is further explored in Lesson 7.3 on page 415.

TEACHING TIP Point out that the proof in Example 2 on page 405 is for just one case of the Reflection Theorem. If students were asked to prove this theorem, they would need to show a proof for each case. They are asked to prove the remaining cases in Exercises 33–35 on page 408.

LESSON 7.3

TEACHING TIP Prior to Activity 7.3 remind students that A' is read as "A prime" **and** the new notation A'' is read as "A double prime." In mapping notation, A'' is usually the image of A'. This notation is also used in Theorem 7.3 and Example 3 on page 414.

TEACHING TIP Be aware that rotational symmetry is defined in this text as a clockwise rotation of 180° or less. Students may ask about counter-clockwise rotations or about clockwise rotations of more than 180° that map a figure onto itself. Also consider discussing what happens with a rotation of 360°.

LESSON 7.4

COMMON ERROR The coordinate rule for a translation might be confusing to students when the values of a or b are negative. Students should think of the positive value of a indicating movement horizontally to the right and a negative value indicating movement horizontally to the left. Similarly, the movement for b is vertical, going up for a positive value and going down for a negative value.

COMMON ERROR Vector notation is new to most students. They may inadvertently use the symbol for a ray instead of the vector symbol since it is more familiar to them. Students may also forget the special brackets and use parentheses. Caution them to check their writing of the notation as they do homework, quizzes, and tests.

LESSON 7.5

TEACHING TIP In addition to Theorem 7.6, the Composition Theorem, on page 431, students may realize that a glide reflection is an isometry because it is a transformation that preserves lengths. There are four isometries: reflection, rotation, translation, and, now, glide reflection. You should be aware that any composition of isometries could be simplified and expressed in terms of one of these four transformations.

Chapter Support

Tips for New Teachers

For use with Chapter 7

COMMON ERROR The composition of a translation and a reflection parallel to the direction of the translation is commutative. When students observe this they tend to think any composition is commutative. They may also think that any composition involving a translation and any reflection is commutative, as well. Use Examples 2 and 3 on page 431 to discuss that not all compositions are commutative and refer to the Study Tip on that page as a reminder.

LESSON 7.6

INCLUSION Ask students to identify and classify frieze patterns in the classroom. Patterns may appear in the clothing they are wearing. You may even wish to wear clothing with a particular frieze pattern for this lesson. You can have other examples available for discussion such as wrapping paper or wallpaper samples. Students with limited English proficiency can benefit from seeing and touching some specific examples in the classroom.

TEACHING TIP Classifying a frieze pattern may be easier for students if you allow them to use the summary table on page 438. Consider having an enlarged copy of the summary table on the bulletin board or perhaps displaying it as an overhead projection for testing or quizzing situations.

Outside Resources

BOOKS/PERIODICALS

Reif, Daniel K. "Architecture and Mathematics." *Mathematics Teacher* (September 1996); pp. 456–458.

SOFTWARE

Perspective Drawing with The Geometer's Sketchpad. Blackline masters and Macintosh and Windows disks with sample sketches and scripts. Berkeley, CA. Key Curriculum Press.

ACTIVITIES/MANIPULATIVES

KaleidoMania! Software and activities that can be used to explore the mathematics of symmetry. Emeryville, CA. Key Curriculum Press.

Johnson, Art, and Joan D. Martin. "The Secret of Anamorphic Art." *Mathematics Teacher* (January 1998); pp. 24–32.

Parent Guide for Student Success

For use with Chapter 7

Chapter Overview One way that you can help your student succeed in Chapter 7 is by discussing the lesson goals in the chart below. When a lesson is completed, ask your student to interpret the lesson goals for you and to explain how the mathematics of the lesson relates to one of the key applications listed in the chart.

Lesson Title	Lesson Goals	Key Applications
7.1: Rigid Motion in a Plane	Identify the three basic rigid transformations. Use transformations in real-life situations.	• Building a Kayak • Stenciling • Machine Embroidery
7.2: Reflections	Identify and use reflections in a plane. Identify relationships between reflections and line symmetry.	• Kaleidoscopes • Delivering Pizza • Molecular Chemistry
7.3: Rotations	Identify rotations in a plane. Use rotational symmetry in real-life situations.	• Logo Design • Wheel Hubs • Escher Art
7.4: Translations and Vectors	Identify and use translations in the plane. Use vectors in real-life situations.	• Boat Navigation • Window Frames • Hot-Air Balloons
7.5: Glide Reflections and Compositions	Identify glide reflections in a plane. Represent transformations as compositions of simpler transformations.	• Pentomino Puzzles • Clothing Patterns • Architecture
7.6: Frieze Patterns	Use transformations to classify frieze patterns. Use frieze patterns to design border patterns in real-life.	• Snakes • Sweater Patterns • Pet Collars

Test-Taking Strategy

*Check your solution another way. If appropriate, **use a visual representation** such as a graph or a diagram as an alternative approach.* Using a different method to check solutions can help your student avoid making the same mistake twice. Encourage your student to sketch graphs or figures or find other methods to check test answers. You may wish to have your student show you an example from the chapter of how to check using a different approach than was used to solve the problem.

NAME _____ DATE _____

Parent Guide for Student Success

For use with Chapter 7

Key Ideas Your student can demonstrate understanding of key concepts by working through the following exercises with you.

Lesson	Exercise
7.1	A pattern for a skirt has one piece that represents half of the front of the skirt. When the front of the skirt is cut out, the fabric is folded and the straight edge of this pattern piece is placed along the folded edge. What type of transformation is being used in creating the front of the skirt?
7.2	The vertices of $\triangle ABC$ are $A(1, 3)$, $B(-1, 1)$, and $C(-3, 2)$. Find the coordinates of the vertices of the reflection of $\triangle ABC$ in the x-axis.
7.3	A circular clockface with no numbers, but with 12 congruent segments marking the hours, has rotational symmetry. What is the smallest angle of rotation?
7.4	The component form of $\overrightarrow{PQ}$ is $\langle 2, -4 \rangle$. Use $\overrightarrow{PQ}$ to translate the triangle with vertices $A(1, 3)$, $B(-1, 1)$, and $C(-3, 2)$.
7.5	Plot the triangle with vertices $A(-2, 5)$, $B(-4, 2)$, and $C(-2, 2)$ on a coordinate plane. Let A', B', and C' be the images of the vertices under a rotation of 180° about the origin. Give the coordinates of A', B', and C'. $\triangle A'B'C'$ is also the result of a composition of two reflections of $\triangle ABC$. Describe the reflections. Does it matter in which order you perform them?
7.6	Classify the frieze pattern.

Home Involvement Activity

You Will Need: Art supplies

Directions: Choose one of the seven categories of frieze patterns and decide on a basic shape to use. Work together to create a frieze pattern. Produce the pattern artistically. You may want to make a stencil (a potato works well for this) and reproduce the pattern using colorful paints. Another option is to press the pattern into clay. Display your frieze pattern in your home.

Answers

7.1: reflection 7.2: $A'(1, -3)$, $B'(-1, -1)$, $C'(-3, -2)$ 7.3: 30°
7.4: $A'(3, -1)$, $B'(1, -3)$, $C'(-1, -2)$ 7.5: $A'(2, -5)$, $B'(4, -2)$, $C'(2, -2)$; a reflection in the x-axis and in the y-axis; no 7.6: TV

NAME _____   DATE _____

Prerequisite Skills Review

For use before Chapter 7

EXAMPLE 1 *Identifying congruent segments*

Use the distance formula to decide whether $\overline{AB} \cong \overline{BC}$.

a. $A(4, -3), B(2, 7), C(6, 1)$ **b.** $A(-3, 5), B(6, 0), C(1, 9)$

SOLUTION

a. $AB = \sqrt{(4-2)^2 + (-3-7)^2}$ **b.** $AB = \sqrt{(-3-6)^2 + (5-0)^2}$

$\quad\quad = \sqrt{2^2 + (-10)^2}$ $\quad\quad = \sqrt{(-9)^2 + 5^2}$

$\quad\quad = \sqrt{4 + 100}$ $\quad\quad = \sqrt{81 + 25}$

$\quad\quad = \sqrt{104}$ $\quad\quad = \sqrt{106}$

$\quad BC = \sqrt{(2-6)^2 + (7-1)^2}$ $\quad BC = \sqrt{(6-1)^2 + (0-9)^2}$

$\quad\quad = \sqrt{(-4)^2 + 6^2}$ $\quad\quad = \sqrt{5^2 + (-9)^2}$

$\quad\quad = \sqrt{16 + 36}$ $\quad\quad = \sqrt{25 + 81}$

$\quad\quad = \sqrt{52}$ $\quad\quad = \sqrt{106}$

$\quad \overline{AB} \not\cong \overline{BC}.$ $\quad \overline{AB} \cong \overline{BC}$

Exercises for Example 1

Use the distance formula to decide whether $\overline{AB} \cong \overline{BC}$.

1. $A(7, 3), B(5, 5), C(6, 8)$ **2.** $A(-2, 1), B(5, -3), C(12, -7)$

3. $A(0, 4), B(6, -4), C(6, 6)$ **4.** $A(-8, 2), B(-4, 11), C(6, 10)$

5. $A(-4, 2), B(-3, 5), C(-2, 8)$ **6.** $A(-2, 7), B(4, 2), C(1, 6)$

EXAMPLE 2 *Identifying corresponding parts of congruent triangles*

Complete the statement, given that $\triangle ABC \cong \triangle XYZ$.

a. $\overline{AC} = \underline{\quad ? \quad}$ **b.** $m \angle Z = \underline{\quad ? \quad}$

SOLUTION

a. $\overline{XZ}$

b. $30°$

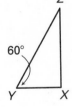

Geometr
Chapter 7 Resource Boo

NAME _____ DATE _____

Prerequisite Skills Review

For use before Chapter 7

Exercises for Example 2

Complete the statement given that $\triangle JKL \cong \triangle MNO$.

7. $\overline{JK} \cong$ ____?____

8. $m \angle K =$ ____?____

9. $m \angle O =$ ____?____

10. $\overline{LK} \cong$ ____?____

11. $JL =$ ____?____

12. $\angle M \cong$ ____?____

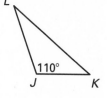
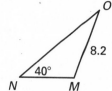

EXAMPLE 3 *Identifying congruent angles and their measure*

Complete the statement given that $l \parallel m$.

a. $m \angle 4 =$ ___?___   **b.** $m \angle 5 =$ ___?___

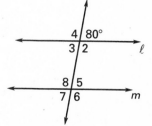

SOLUTION

a. $m \angle 4 = 100$ because $m \angle 1$ and $m \angle 4$ are supplementary.

b. $m \angle 5 = 80°$ because $\angle 5 \cong \angle 1$ by the Corresponding Angles Theorem.

Exercises for Example 3

Complete the statement given that $l \parallel m$.

13. $m \angle 8 =$ ___?___

14. $m \angle 4 =$ ___?___

15. $m \angle 5 =$ ___?___

16. $m \angle 7 =$ ___?___

17. $m \angle 2 + m \angle 6 =$ ___?___

18. $m \angle 3 =$ ___?___

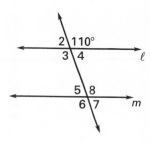

Strategies for Reading Mathematics

For use with Chapter 7

Strategy: Reading Graphs of Transformations

Transformations are often shown on a coordinate plane. You need to know how to read the graphs of transformed figures so you can identify the corresponding parts of the figures.

In the graph at the right, △MNP has been reflected in the x-axis. The image of △MNP is △M′N′P′. As with congruent figures, corresponding vertices are listed in the same order for each figure. You can also say that △MNP is the preimage of △M′N′P′.

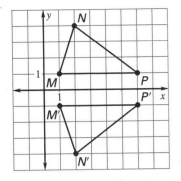

△MNP → △M′N′P′
means
triangle MNP is
mapped onto
triangle M′N′P′.

STUDY TIP

Reading Transformed Points

To identify the image of a point, look for a point with the same name followed by a prime symbol. For instance, if you see point R, look for R′ to find its image.

Questions

1. Name the image of $\overline{MN}$. Explain how you know your answer is correct.

2. Name the image of $\overline{NP}$. Explain how you know your answer is correct.

3. Name the preimage of $\overline{P'M'}$. Explain how you know your answer is correct.

4. Use arrow notation to describe the transformation shown at the right.

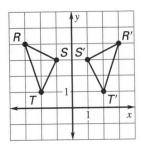

Strategies for Reading Mathematics

For use with Chapter 7

Visual Glossary

The Study Guide on page 394 lists the key vocabulary for Chapter 7 as well as review vocabulary from previous chapters. Use the page references on page 394 or the Glossary in the textbook to review key terms from prior chapters. Use the visual glossary below to help you understand some of the key vocabulary in Chapter 7. You may want to copy these diagrams into your notebook and refer to them as you complete the chapter.

GLOSSARY

image (p. 396) The new figure that results from the transformation of a figure in a plane.

preimage (p. 396) The original figure in the transformation of a figure in a plane.

transformation (p. 396) The operation that maps, or moves, a preimage onto an image.

isometry (p. 397) A transformation that preserves lengths.

reflection (p. 404) A type of transformation that uses a line that acts like a mirror, called the line of reflection, with an image reflected in the line.

rotation (p. 412) A type of transformation in which a figure is turned about a fixed point, called the center of rotation.

translation (p. 421) A type of transformation that maps every two points P and Q in the plane to points P' and Q', so that the following two properties are true. (1) $PP' = QQ'$. (2) $\overline{PP'} \parallel \overline{QQ'}$, or $\overline{PP'}$ and $\overline{QQ'}$ are collinear.

Transforming Geometric Figures

You can transform figures in a plane to produce new figures. Every transformation involves a preimage and an image. Some transformations are isometries, and are also called rigid transformations.

preimage   image

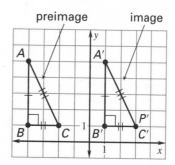

The transformation is an isometry.

Three Basic Transformations

Three basic transformations are reflections, rotations, and translations.

Reflection

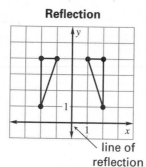

line of reflection

Rotation

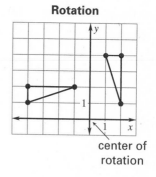

center of rotation

Translation

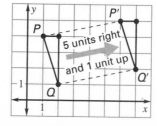

Segments PP' and QQ' are congruent and parallel.

TEACHER'S NAME _____ CLASS _____ ROOM _____ DATE _____

Lesson Plan

2-day lesson (See *Pacing the Chapter,* TE pages 392C–392D)   For use with pages 395–402

GOALS
1. **Identify the three basic rigid transformations.**
2. **Use transformations in real-life situations.**

State/Local Objectives _____

_____

✓ **Check the items you wish to use for this lesson.**

STARTING OPTIONS
____ Prerequisite Skills Review: CRB pages 5–6
____ Strategies for Reading Mathematics: CRB pages 7–8
____ Homework Check: TE page 376: Answer Transparencies
____ Warm-Up or Daily Homework Quiz: TE pages 396 and 380, CRB page 11, or Transparencies

TEACHING OPTIONS
____ Motivating the Lesson: TE page 397
____ Concept Activity: SE page 395
____ Lesson Opener (Geometry Software): CRB page 12 or Transparencies
____ Examples: Day 1: 1–5, SE pages 396–398; Day 2: See the Extra Examples.
____ Extra Examples: Day 1 or Day 2: 1–5, TE pages 397–398 or Transp.
____ Closure Question: TE page 398
____ Guided Practice: SE page 399   Day 1: Exs. 1–11; Day 2: See Checkpoint Exs. TE pages 397–398

APPLY/HOMEWORK
Homework Assignment
____ Basic   Day 1: 12–42 even; Day 2: 13–39 odd, 44, 45, 47–58
____ Average   Day 1: 12–42 even; Day 2: 13–39 odd, 44, 45, 47–58
____ Advanced   Day 1: 12–42 even; Day 2: 13–39 odd, 44–58

Reteaching the Lesson
____ Practice Masters: CRB pages 13–15 (Level A, Level B, Level C)
____ Reteaching with Practice: CRB pages 16–17 or Practice Workbook with Examples
____ Personal Student Tutor

Extending the Lesson
____ Applications (Real-Life): CRB page 19
____ Challenge: SE page 402; CRB page 20 or Internet

ASSESSMENT OPTIONS
____ Checkpoint Exercises: Day 1 or Day 2: TE pages 397–398 or Transp.
____ Daily Homework Quiz (7.1): TE page 402, CRB page 23, or Transparencies
____ Standardized Test Practice: SE page 402; TE page 402; STP Workbook; Transparencies

Notes _____

_____

_____

TEACHER'S NAME _____ CLASS _____ ROOM _____ DATE _____

Lesson Plan for Block Scheduling

1-day lesson (See *Pacing the Chapter,* TE pages 392C–392D)   For use with pages 395–402

GOALS   1. **Identify the three basic rigid transformations.**
            2. **Use transformations in real-life situations.**

State/Local Objectives _____

_____

_____

✓ Check the items you wish to use for this lesson.

STARTING OPTIONS
____ Prerequisite Skills Review: CRB pages 5–6
____ Strategies for Reading Mathematics: CRB pages 7–8
____ Homework Check: TE page 376: Answer Transparencies
____ Warm-Up or Daily Homework Quiz: TE pages 396 and
        380, CRB page 11, or Transparencies

TEACHING OPTIONS
____ Motivating the Lesson: TE page 397
____ Concept Activity: SE page 395
____ Lesson Opener (Geometry Software): CRB page 12 or Transparencies
____ Examples 1–5: SE pages 396–398
____ Extra Examples 1–5: TE pages 397–398 or Transparencies
____ Closure Question: TE page 398
____ Guided Practice Exercises: SE page 399

APPLY/HOMEWORK
Homework Assignment
____ Block Schedule: 12–42, 44, 45, 47–58

Reteaching the Lesson
____ Practice Masters: CRB pages 13–15 (Level A, Level B, Level C)
____ Reteaching with Practice: CRB pages 16–17 or Practice Workbook with Examples
____ Personal Student Tutor

Extending the Lesson
____ Applications (Real-Life): CRB page 19
____ Challenge: SE page 402; CRB page 20 or Internet

ASSESSMENT OPTIONS
____ Checkpoint Exercises: TE pages 397–398 or Transparencies
____ Daily Homework Quiz (7.1): TE page 402, CRB page 23, or Transparencies
____ Standardized Test Practice: SE page 402; TE page 402; STP Workbook; Transparencies

CHAPTER PACING GUIDE	
Day	**Lesson**
1	**7.1 (all)**
2	7.2 (all)
3	7.3 (all)
4	7.4 (all); 7.5 (begin)
5	7.5 (end); 7.6 (begin)
6	7.6 (end); Review Ch. 7
7	Assess Ch. 7; 8.1 (all)

Notes _____

_____

_____

NAME _____ DATE _____

WARM-UP EXERCISES

For use before Lesson 7.1, pages 395–402

1. $\triangle XYZ \cong \triangle MNO$. List the pairs of corresponding sides and angles.

2. Quadrilateral $ABCD \cong$ quadrilateral $PQRS$. The measure of $\angle A = 62°$, $m\angle B = 128°$, $m\angle C = 97°$. Find $m\angle S$.

3. Name 3 points the same distance from $(0, 0)$ as $(2, 5)$.

4. If $B(-1, 4)$ is moved right 2 units and down 1 unit, where is its new location?

DAILY HOMEWORK QUIZ

For use after Lesson 6.7, pages 371–380

Find the area of the polygon.

1.

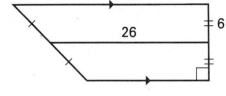

26
6

2.

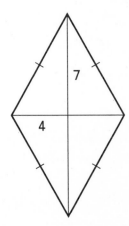

7
4

3.

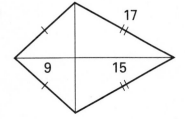

17
9   15

4.

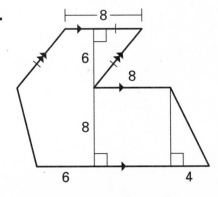

8
6
8
8
6   4

NAME _____ DATE _____

Geometry Software Lesson Opener

For use with pages 396–402

Use geometry software to experiment with three basic transformations–translations, reflections, and rotations. Before you begin, select the option to keep the preimage displayed.

1. Draw a square with a side length of 1 in. Translate the square three times: (1) 1 inch horizontally and 0 inches vertically; (2) 0 inches horizontally and 1 inch vertically; (3) −1 inch horizontally and 0 inches vertically. On paper, sketch your final figure and shade the original square.

2. Draw a square with a side length of 1 inch. Use *translations* to draw a rectangle as shown at the right. Explain your steps.

3. Draw a right triangle ABC with right angle at C. Reflect $\triangle ABC$ in $\overline{BC}$. Label the new vertex A'. Reflect $\triangle A'BC$ in $\overline{CA'}$. Label the new vertex B'. Reflect $\triangle A'B'C$ in $\overline{CB'}$. On paper, sketch your final figure and shade $\triangle ABC$. Name the shape of figure $ABA'B'$.

4. Draw a scalene obtuse triangle ABC. Mark A as the center of rotation and rotate $\triangle ABC$ 180°. Label the new vertices B' and C'. Construct segments $\overline{BC'}$ and $\overline{CB'}$. On paper, sketch your final figure and shade $\triangle ABC$. Name the shape of figure $BCB'C'$.

Geometry
Chapter 7 Resource Book

Practice A

For use with pages 396–402

Name the transformation that maps the lighter checkered flag (preimage) onto the darker checkered flag (image).

1.

2.

3.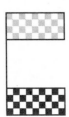

Decide whether the transformation is an isometry. If it is, name the transformation. (Preimages are unshaded; images are shaded.)

4.

5.

6.

Use the graph of the transformation below. *ABCDE* is the preimage.

7. Figure *ABCDE* → Figure _____.

8. Name and describe the transformation.

9. Name the image of $\overline{CD}$.

10. Name the preimage of $\overline{FJ}$.

11. Name the coordinates of the preimage of point *I*.

12. Show that $\overline{DE}$ and $\overline{IJ}$ have the same length, using the Distance Formula.

Use the diagrams to complete the statement.

13. △*ABC* → △___?___

14. △*DEF* → △___?___

15. △___?___ → △*ACB*

16. △___?___ → △*CBA*

17. △*RQP* → △___?___

18. △___?___ → △*EFD*

Sketch the preimage if the image was transformed by the following.

19. Reflection

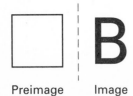

Preimage   Image

20. Rotation of 180° clockwise

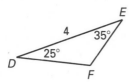

Preimage   Image

21. A non-rigid transformation

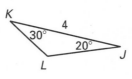

Preimage   Image

Name the transformation that maps the unshaded turtle (preimage) onto the shaded turtle (image).

1.

2.

3.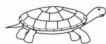

Use the graph of the transformation below. *ABCDEF* is the preimage.

4. Figure *ABCDEF* → Figure _____.

5. Name and describe the transformation.

6. Name the image of $\overline{CD}$.

7. Name the preimage of $\overline{HI}$.

8. Name the coordinates of the preimage of point *J*.

9. Show that $\overline{EF}$ and $\overline{KL}$ have the same length, using the Distance Formula.

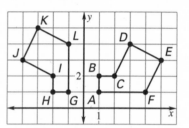

In Exercises 10–13, name the transformation that will map Tree *A* onto the indicated tree.

10. Tree *B*   **11.** Tree *C*

12. Tree *D*   **13.** Tree *E*

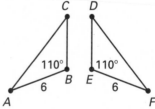

In Exercises 14–15, complete the statement regarding the transformation shown.

14. △*ABC* → △____?____

15. △____?____ → △*EDF*

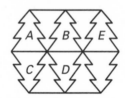

Find the value of each variable, given that the transformation is an isometry.

16.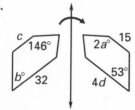

17.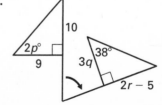

NAME _____ DATE _____

Practice C

For use with pages 396–402

Name the transformation that maps the unshaded arrow (preimage) onto the shaded arrow (image).

1.

2.

3.

Use the graph of the transformation below. *ABCD* **is the preimage.**

4. Figure $ABCD \rightarrow$ Figure _____.

5. Name and describe the transformation.

6. Name the image of $\overline{CD}$.

7. Name the preimage of $\overline{FG}$.

8. Name the coordinates of the preimage of point *H*.

9. Sketch the image that results when figure *ABCD* is reflected over the *x*-axis followed by a reflection over the *y*-axis.

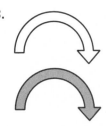

Show that the transformation is an isometry by using the Distance Formula to compare the side lengths of the triangles.

10. $\triangle ABC \rightarrow \triangle DEF$

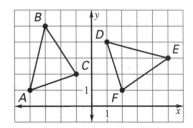

11. $\triangle PQR \rightarrow \triangle MNO$

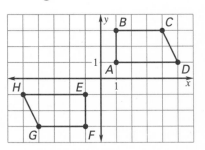

Find the value of each variable, given that the transformation is an isometry.

12.

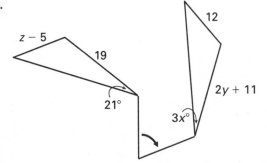

13.

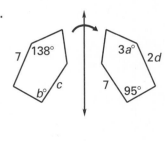

NAME _____ DATE _____

Reteaching with Practice

For use with pages 396–402

GOAL Identify the three basic rigid transformations.

VOCABULARY

Figures in a plane can be reflected, rotated, or translated to produce new figures. The new figure is called the **image,** and the original figure is called the **preimage.**

The operation that *maps*, or moves, the preimage onto the image is called a **transformation.**

An **isometry** is a transformation that preserves lengths.

EXAMPLE 1 *Naming Transformations*

Use the graph of the transformation at the right.

a. Name and describe the transformation.

b. Name the coordinates of the vertices of the image.

c. Is △ABC congruent to its image?

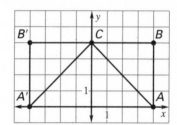

SOLUTION

a. The transformation is a reflection in the y-axis.

b. The coordinates of the vertices of the image, △A′B′C, are A′(−4, 0), B′(−4, 4), and C(0, 4).

c. Yes, △ABC is congruent to its image △A′B′C. One way to show this would be to use the Distance Formula to find the lengths of the sides of both triangles. Then use the SSS Congruence Postulate.

Exercises for Example 1

In Exercises 1–3, use the graph of the transformation to answer the questions.

1. Name and describe the transformation.

2. Name the coordinates of the vertices of the image.

3. Name two angles with the same measure.

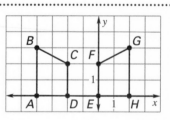

EXAMPLE 2 *Identifying Isometries*

Which of the following transformations appear to be isometries?

a.

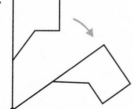

b.

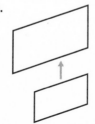

c.

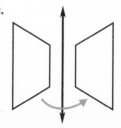

NAME _____ DATE _____

Reteaching with Practice

For use with pages 396–402

SOLUTION

a. This transformation appears to be an isometry. The image on the left is rotated about a point to produce a congruent image on the right.

b. This transformation is not an isometry. The parallelogram on the top is not congruent to its preimage on the bottom.

c. This transformation appears to be an isometry. The trapezoid on the left is reflected in a line to produce a congruent trapezoid on the right.

Exercises for Example 2

State whether the transformation appears to be an isometry.

4.

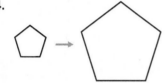

5.

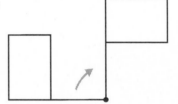

EXAMPLE 3 *Preserving Length and Angle Measure*

In the diagram, *ABCD* is mapped onto *WXYZ*. The mapping is a reflection. Given that *ABCD* → *WXYZ* is an isometry, find the length of $\overline{WX}$ and the measure of ∠*Y*.

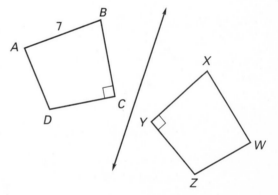

SOLUTION

Because the transformation is an isometry, the two figures are congruent. So, *WX* = *AB* = 7 and *m*∠*Y* = *m*∠*C* = 90°.

Exercises for Example 3

In Exercises 6 and 7, find the value of each variable, given that the transformation is an isometry.

6.

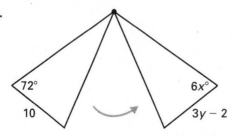

7.

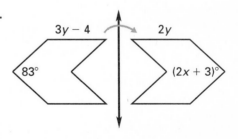

NAME _____ DATE _____

Quick Catch-Up for Absent Students

For use with pages 395–402

The items checked below were covered in class on (date missed) _____

Activity 7.1: Motion in the Plane (p. 395)

_____ **Goal:** Determine which types of motion in a plane maintain the congruence of a figure.

Lesson 7.1: Rigid Motion in a Plane

_____ **Goal 1:** Identify the three basic rigid transformations. (pp. 396–397)

Material Covered:

_____ Example 1: Naming Transformations

_____ Student Help: Study Tip

_____ Example 2: Identifying Isometries

_____ Example 3: Preserving Length and Angle Measure

Vocabulary:

image, p. 396   preimage, p. 396
transformation, p. 396   isometry, p. 397

_____ **Goal 2:** Use transformations in real-life situations. (p. 398)

Material Covered:

_____ Example 4: Identifying Transformations

_____ Example 5: Using Transformations

_____ Other (specify) _____

_____

Homework and Additional Learning Support

_____ Textbook (specify) _pp. 399–402_____

_____

_____ *Reteaching with Practice* worksheet (specify exercises)_____

_____ *Personal Student Tutor* for Lesson 7.1

NAME _____ DATE _____

Real-Life Application: When Will I Ever Use This?

For use with pages 396–402

Braille

A variety of things, such as injuries, diseases, or lesions of the brain or optic nerve may cause blindness, or the inability to see. This physical impairment may make everyday living a challenge. However, inventions such as the writing system called Braille, has made communicating for the blind easier.

Braille is not a language, rather it is a code by which languages may be written or read by the blind. It consists of a series of cells with raised dots representing the letters of the alphabet. A full Braille cell has six raised dots arranged in two parallel columns, each having three dots. Sixty-four combinations of raised dots are possible in each cell. Braille is read by moving the hands over the raised dots from left to right along each line. The average reading speed is about 125 words per minute.

Louis Braille invented the system of Braille in 1824. After losing his sight at the age of three as a result of an accident, he attended the National Institute for Blind Youth in Paris, France, where, as a student he yearned for books to read. He invented a writing system at the age of 15 that evolved from the night writing code invented by Charles Barbier for sending military messages that could be read at night without light.

In Exercises 1 and 2, use the following information.

The first ten Braille letters use dots in a 2×2 matrix. The next 16 letters use dots in a 3×2 matrix.

1. Which Braille letters have rotations that are different Braille letters? Describe the rotation.

2. Which Braille letters have reflections that are different Braille letters? Describe the reflection.

NAME _____ DATE _____

Challenge: Skills and Applications

For use with pages 396–402

1. Write a two-column proof to show that an isometry preserves "betweenness." (This, in turn, shows that an isometry preserves lines.)

    Given: Q is between P and R, $PQ = P'Q'$, $QR = Q'R'$, $PR = P'R'$

    Prove: Q' is between P' and R'.

2. Write a paragraph proof to show that an isometry preserves angles.

    Given: $\angle PQR$ is an angle, $PQ = P'Q'$, $QR = Q'R'$, $PR = P'R'$

    Prove: $\angle PQR \cong \angle P'Q'R'$

In Exercises 3–6, a transformation and a preimage point are given. Use the following information to find the image point.

In a coordinate plane, a transformation can be described using algebraic notation. For example, $(x, y) \rightarrow (x + 2, 5 - y)$ means "to get the coordinates of the image point, add 2 to the x-coordinate of the preimage, and subtract the y-coordinate of the preimage from 5." Under this transformation, the preimage point $(4, 7)$ would correspond to the image point $(4 + 2, 5 - 7)$, or $(6, -2)$.

3. preimage: $(5, -4)$; transformation: $(x, y) \rightarrow (8 - x, y + 3)$

4. preimage: $(3, 7)$; transformation: $(x, y) \rightarrow (2x, 3y)$

5. preimage: $(-8, -4)$; transformation: $(x, y) \rightarrow (-x, y)$

6. preimage: $(9, 1)$; transformation: $(x, y) \rightarrow (-y, x)$

In Exercises 7–12, sketch the image of the given triangle after the given transformation. Give the coordinates of the image. Based on your graph, identify the transformation as a reflection, rotation, translation, or other transformation, and tell whether the transformation appears to be an isometry.

7. $(x, y) \rightarrow (x + 4, y - 3)$

8. $(x, y) \rightarrow (2x, 2y)$

9. $(x, y) \rightarrow (x, -y)$

10. $(x, y) \rightarrow (y, -x)$

11. $(x, y) \rightarrow (5 - x, y)$

12. $(x, y) \rightarrow (x + 4, -y)$

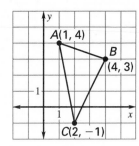

Geometry
Chapter 7 Resource Book

TEACHER'S NAME _____ CLASS _____ ROOM _____ DATE _____

Lesson Plan

2-day lesson (See *Pacing the Chapter,* TE pages 392C–392D)   **For use with pages 403–410**

GOALS   1. **Identify and use reflections in a plane.**
2. **Identify relationships between reflections and line symmetry.**

State/Local Objectives _____

_____

✓ **Check the items you wish to use for this lesson.**

STARTING OPTIONS
____ Homework Check: TE page 399: Answer Transparencies
____ Warm-Up or Daily Homework Quiz: TE pages 404 and 402, CRB page 23, or Transparencies

TEACHING OPTIONS
____ Concept Activity: SE page 403
____ Lesson Opener (Activity): CRB page 24 or Transparencies
____ Technology Activity with Keystrokes: CRB pages 25–28
____ Examples: Day 1: 1, 2, 4, SE pages 404–406; Day 2: 3, 5, SE pages 405–406
____ Extra Examples: Day 1: TE pages 405–406 or Transp.; Day 2: TE pages 405–406 or Transp.
____ Closure Question: TE page 406
____ Guided Practice: SE page 407   Day 1: Exs. 1–11; Day 2: Exs. 12–14

APPLY/HOMEWORK
Homework Assignment
____ Basic   Day 1: 15–35; Day 2: 36–46, 48–51, 57–71
____ Average   Day 1: 15–35; Day 2: 36–46, 48–51, 51–71
____ Advanced   Day 1: 15–35; Day 2: 36–46, 48–71

Reteaching the Lesson
____ Practice Masters: CRB pages 29–31 (Level A, Level B, Level C)
____ Reteaching with Practice: CRB pages 32–33 or Practice Workbook with Examples
____ Personal Student Tutor

Extending the Lesson
____ Applications (Interdisciplinary): CRB page 35
____ Challenge: SE page 410; CRB page 36 or Internet

ASSESSMENT OPTIONS
____ Checkpoint Exercises: Day 1: TE pages 405–406 or Transp.; Day 2: TE pages 405–406 or Transp.
____ Daily Homework Quiz (7.2): TE page 410, CRB page 39, or Transparencies
____ Standardized Test Practice: SE page 410; TE page 410; STP Workbook; Transparencies

Notes _____

_____

_____

TEACHER'S NAME _____ CLASS _____ ROOM _____ DATE _____

Lesson Plan for Block Scheduling
1-day lesson (See *Pacing the Chapter*, TE pages 392C–392D)   For use with pages 403–410

GOALS   1. **Identify and use reflections in a plane.**
2. **Identify relationships between reflections and line symmetry.**

State/Local Objectives _____

_____

_____

✓ **Check the items you wish to use for this lesson.**

STARTING OPTIONS
____ Homework Check: TE page 399: Answer Transparencies
____ Warm-Up or Daily Homework Quiz: TE pages 404 and
          402, CRB page 23, or Transparencies

TEACHING OPTIONS
____ Concept Activity: SE page 403
____ Lesson Opener (Activity): CRB page 24 or Transparencies
____ Technology Activity with Keystrokes: CRB pages 25–28
____ Examples 1–5: SE pages 404–406
____ Extra Examples: TE pages 405–406 or Transparencies
____ Closure Question: TE page 406
____ Guided Practice Exercises: SE page 407

APPLY/HOMEWORK
Homework Assignment
____ Block Schedule: 15–46, 48–51, 57–71

Reteaching the Lesson
____ Practice Masters: CRB pages 29–31 (Level A, Level B, Level C)
____ Reteaching with Practice: CRB pages 32–33 or Practice Workbook with Examples
____ Personal Student Tutor

Extending the Lesson
____ Applications (Interdisciplinary): CRB page 35
____ Challenge: SE page 410; CRB page 36 or Internet

ASSESSMENT OPTIONS
____ Checkpoint Exercises: TE pages 405–406 or Transparencies
____ Daily Homework Quiz (7.2): TE page 410, CRB page 39, or Transparencies
____ Standardized Test Practice: SE page 410; TE page 410; STP Workbook; Transparencies

CHAPTER PACING GUIDE	
Day	Lesson
1	7.1 (all)
2	**7.2 (all)**
3	7.3 (all)
4	7.4 (all); 7.5 (begin)
5	7.5 (end); 7.6 (begin)
6	7.6 (end); Review Ch. 7
7	Assess Ch. 7; 8.1 (all)

Notes _____

_____

_____

Lesson 7.2

WARM-UP EXERCISES

For use before Lesson 7.2, pages 403–410

1. $P(5, 4)$ is reflected in the *x*-axis. In which quadrant is its image?

2. $P(5, 4)$ is reflected in the *y*-axis. In which quadrant is its image?

3. What is the hypothesis of the statement *If Q lies on m and* $\overline{PQ} \perp m$, *then* $\overline{PQ} \cong \overline{P'Q'}$?

4. If $\angle A \cong \angle A'$, $\angle B \cong \angle B'$, and $\overline{AB} \cong \overline{A'B'}$, by what congruence postulate is $\triangle ABC \cong \triangle A'B'C'$?

....

DAILY HOMEWORK QUIZ

For use after Lesson 7.1, pages 395–402

1. $ABCD \rightarrow$ _____ .

2. Name the transformation.

3. Name the coordinates of the image of point *D*.

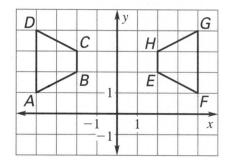

4. Is the transformation an isometry? Explain.

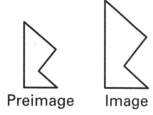

Preimage   Image

5. Find the value of each variable, given that the transformation is an isometry.

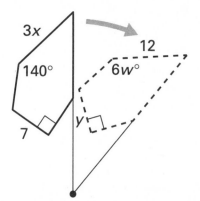

SET UP: Work with a partner. If possible, choose a partner who is about the same height as you are.

YOU WILL NEED: • masking tape

1. Place a 5-ft long strip of masking tape on the floor. Imagine a large mirror that extends up from the tape. Stand a few feet apart from your partner, facing your partner and the imaginary mirror. Decide which one of you will be the preimage. The other will be the image after a reflection in the mirror. To begin, the preimage assumes a position such as arms outstretched. The image then mirrors this position. Repeat for 10 different positions, then switch roles for 10 different positions. If the preimage extends a left arm up, what does the image do?

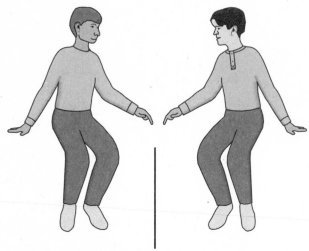

2. Now stand side-by-side with your partner with the imaginary mirror between you. Again the preimage assumes positions that the image mirrors. If the preimage places its closer hand "on" the mirror, what does the image do?

3. Now imagine that the tape is a mirror *line* on the floor, rather than a plane. When the preimage assumes a position, the two feet on the floor act as points that are reflected in the line, resulting in two image points. On a coordinate plane, draw and label all four points for one of your positions and its image, using the *x*-axis or *y*-axis as the mirror line.

NAME _____ DATE _____

Technology Activity

For use with pages 404–410

GOAL **To use geometry software to verify statements about reflections in a plane**

Geometry software can be used to verify statements about reflections in a plane. For example, you could use geometry software to construct $\overline{AB}$. Then, you could use the software's tools to reflect $\overline{AB}$ in a line to produce $\overline{A'B'}$. You could then verify that the reflection preserved the length of the original segment.

Given: A reflection in m maps A onto A' and B onto B'.

Prove: $AB = A'B'$

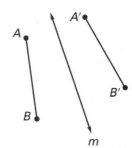

Activity

① Draw $\overline{AB}$.

② Draw line m such that line m does not intersect $\overline{AB}$ (see figure above).

③ Use the software's reflection feature to reflect $\overline{AB}$ in line m.

④ Measure the lengths of $\overline{AB}$ and $\overline{A'B'}$.

Exercises

In Exercises 1–3, a reflection in m maps A onto A' and B onto B'. Use geometry software to verify $AB = A'B'$.

1.

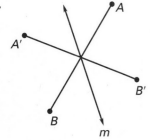

2. A lies on m.

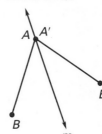

3. A lies on m, $\overline{AB} \perp m$.

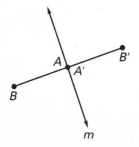

Technology Activity Keystrokes

For use with pages 404–410

TI-92

1. Draw $\overline{AB}$.

   F2 5 (Move cursor to location for point A.) **ENTER** A (Move cursor to location for point B.) **ENTER** B

2. Draw line m such that line m does not intersect $\overline{AB}$.

   F2 4 **ENTER** m (Move cursor and draw line so that it does not intersect $\overline{AB}$.)

   ENTER

3. Use the software's reflection feature to reflect $\overline{AB}$ in line m.

   F5 4 (Move cursor to $\overline{AB}$.) **ENTER** (Move cursor to line m.) **ENTER**

4. Measure the lengths of $\overline{AB}$ and $\overline{A'B'}$.

   F6 1 (Move cursor to $\overline{AB}$.) **ENTER** (Move cursor to $\overline{A'B'}$.) **ENTER**

SKETCHPAD

1. Draw $\overline{AB}$ using the segment straightedge tool.

2. Choose the line straightedge tool and draw line m such that line m does not intersect $\overline{AB}$.

3. Choose the selection arrow tool and select line m. Choose **Mark Mirror** from the **Transform** menu.

4. Use the selection arrow tool to select A, B, and $\overline{AB}$. Choose **Reflect** from the **Transform** menu.

5. Choose the selection arrow tool and select $\overline{AB}$. Then hold down the shift key, select $\overline{A'B'}$, and choose **Length** from the **Measure** menu.

NAME _____ DATE _____

Technology Activity Keystrokes

For use with page 409

Keystrokes for Exercise 47
TI-92

1. Draw a polygon.

    F3 4 (Locate first vertex.) **ENTER** (Move to second vertex.) **ENTER**
    (Move to third vertex.) **ENTER** (Repeat for as many vertices as are desired
    and then return to the first vertex to close the polygon.) **ENTER**

2. Draw line *m*.

    F2 4 (Locate position for line *m*.) **ENTER** (Move cursor to place line.)

    ENTER *m*

3. Reflect the polygon in line *m*.

    F5 4 (Place the cursor on the polygon.) **ENTER** (Move the cursor to line *m*.)

    ENTER

4. Draw a segment connecting a vertex of the polygon to its corresponding
    vertex on the reflection.

    F2 5 (Place cursor on a vertex of the polygon.) **ENTER** (Move cursor to the
    corresponding vertex on the reflection.) **ENTER**

5. **F6** 1 (Move cursor to vertex of preimage.) **ENTER** (Move cursor line to *m*.)
    ENTER (Move cursor to corresponding vertex of image.) **ENTER**
    (Move cursor to line *m*.)

    Repeat this process for each vertex.

Technology Activity Keystrokes

For use with page 409

SKETCHPAD

1. Draw a polygon using the segment straightedge tool.

2. Draw line *m*. Choose line from the straightedge tool. Draw the line so that it does not intersect the polygon.

3. Reflect the polygon in line *m* (select the line using the selection arrow tool). Choose **Mark Mirror** from the **Transform** menu. Select the segments and the point of the polygon. Choose **Reflect** from the **Transform** menu.

4. Draw a segment connecting a vertex on the polygon to its corresponding vertex on the reflection using the segment straightedge tool.

5. Measure the distance between a vertex and line *m*. Choose the selection arrow tool and select a vertex on the preimage. Then hold down the shift key, select line *m*, and choose **Distance** from the **Measure** menu. Select the corresponding vertex on the image, hold down the shift key, select line *m*, and choose **Distance** from the **Measure** menu. Compare the results. Repeat this process for the other vertices.

Lesson 7.2

NAME _____ DATE _____

Practice A

For use with pages 404–410

Determine whether the light figure maps onto the darker figure by a reflection in line *m*.

1.
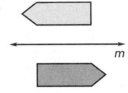

2.
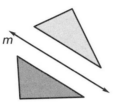

3.

Use the diagram at the right to complete the statement.

4. $\overline{AB} \to$ __?__

5. $\overline{AD} \to$ __?__

6. $\angle D \to$ __?__

7. __?__ $\to \angle BFG$

8. $\angle EBA \to$ __?__

9. $B \to$ __?__

10. figure $ADEB \to$ figure __?__

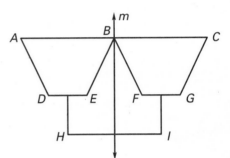

Trace the figure and draw its reflection in the line ℓ.

11.

12.

13.
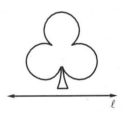

Find the coordinates of the reflection without using a coordinate plane. Then check your answer by plotting the image and preimage on a coordinate plane.

14. $M(4, 2)$ is reflected in the *x*-axis.

15. $N(3, 5)$ is reflected in the *x*-axis.

16. $O(5, 1)$ is reflected in the *y*-axis.

17. $P(3, 0)$ is reflected in the *y*-axis.

Sketch the figure, if possible.

18. A triangle with exactly one line of symmetry

19. A trapezoid with exactly one line of symmetry

20. A pentagon with exactly one line of symmetry

21. A hexagon with exactly two lines of symmetry

NAME _____ DATE _____

Practice B

For use with pages 404–410

Trace the figure and draw its reflection in the line ℓ.

1.

2.

3.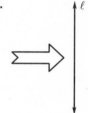

Decide whether the conclusion is _true_ or _false_.

4. If $M(2, 3)$ is reflected in the line $y = 4$, then M' is $(6, 3)$.

5. If $N(-3, 1)$ is reflected in the line $y = -2$, then N' is $(-1, 1)$.

6. If $P(0, -2)$ is reflected in the line $x = 2$, then P' is $(0, 6)$.

7. If $Q(4, -3)$ is reflected in the line $x = 2$, then Q' is $(0, -3)$.

Use the diagram at the right to name the image of $\triangle 1$ after the reflection.

8. Reflection in the x-axis

9. Reflection in the y-axis

10. Reflection in the line $y = x$

11. Reflection in the line $y = -x$

12. Reflection in the y-axis, followed by a reflection in the x-axis

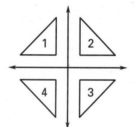

Sketch the figure, if possible.

13. A triangle with exactly two lines of symmetry

14. A quadrilateral with exactly two lines of symmetry

15. A pentagon with exactly two lines of symmetry

16. A hexagon with exactly two lines of symmetry

Use the diagram at the right to answer the following.

17. Underground cable is to be laid so that two new homes may have electricity. Where along the road (line m) should the transformer box be placed so that there is a minimum distance from the box to each of the homes?

18. Measure the minimum distance to the nearest tenth of a centimeter.

NAME _____ DATE _____

Practice C

For use with pages 404–410

Trace the figure and draw its reflection in the line ℓ.

1.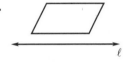

2.

3.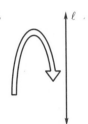

Find the coordinates of the reflection without using a coordinate plane. Then check your answer by plotting the image and preimage on a coordinate plane.

4. $M(3, 4)$ is reflected in the line $y = 1$.

5. $N(-2, 2)$ is reflected in the line $y = -1$.

6. $P(-2, 3)$ is reflected in the line $x = -3$.

7. $Q(5, -2)$ is reflected in the line $x = 3$.

Use the diagram at the right to name the image of Segment 1 after the reflection.

8. Reflection in the x-axis

9. Reflection in the y-axis

10. Reflection in the line $y = x$

11. Reflection in the line $y = -x$

12. Reflection in the y-axis, followed by a reflection in the x-axis

13. Reflection in the x-axis, followed by a reflection in the y-axis

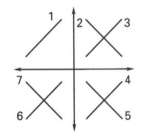

Use the diagram at the right to answer the following.

14. Reflect the word in line m. What does it *read*?

15. What letters of the alphabet have horizontal line symmetry?

16. Use the letters you found in Exercise 14 to write a mirror message.

CHECKBOOK

$\longleftrightarrow m$

Find point *C* on the *x*-axis so *AC* + *BC* is a minimum.

17. $A(1, 4), B(4, -1)$

18. $A(-2, 3), B(-4, 0)$

19. $A(-3, 2), B(-6, -4)$

20. $A(1, -3), B(-1, 1)$

The points *A*(2, 5) and *B*(−4, −7) are reflection images of one another.

21. Find the coordinates of the midpoint of $\overline{AB}$.

22. Find the slope of $\overline{AB}$.

23. Find the slope of a line perpendicular to $\overline{AB}$.

24. Use your answers to Exercises 21 and 23 to write the equation of the line in which A is reflected to B.

NAME _____ DATE _____

Reteaching with Practice

For use with pages 404–410

GOAL **Identify and use reflections in a plane and identify relationships between reflections and line symmetry.**

VOCABULARY

A transformation which uses a line that acts like a mirror, with an image reflected in the line, is called a **reflection.** The line which acts like a mirror in a reflection is called the **line of reflection.**

A figure in the plane has a **line of symmetry** if the figure can be mapped onto itself by a reflection in the line.

Theorem 7.1 Reflection Theorem
A reflection is an isometry.

EXAMPLE 1 *Reflections in a Coordinate Plane*

Graph the given reflection.

a. $A(3, 2)$ in the y-axis

b. $B(1, -3)$ in the line $y = 1$

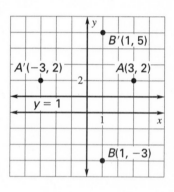

SOLUTION

a. Since A is three units to the right of the y-axis, its reflection, A', is three units to the left of the x-axis.

b. Start by graphing $y = 1$ and B. From the graph, you can see that B is 4 units below the line of reflection. This implies that its reflection, B', is 4 units above the line.

Exercises for Example 1

In Exercises 1–8, graph the given reflection.

1. $C(-1, 4)$ in the x-axis

2. $D(0, 3)$ in the y-axis

3. $E(4, -2)$ in the line $y = 3$

4. $F(1, -2)$ in the line $y = -2$

5. $G(3, 5)$ in the line $x = 1$

6. $H(-3, -1)$ in the line $x = 4$

7. $I(4, 5)$ in the line $x = -2$

8. $J(-2, 3)$ in the line $y = 1$

EXAMPLE 2 *Finding Lines of Symmetry*

Triangles can have different lines of symmetry depending on their shape. Find the number of lines of symmetry a triangle has when it is one of the following.

a. equilateral

b. isosceles

c. scalene

Reteaching with Practice

For use with pages 404–410

SOLUTION

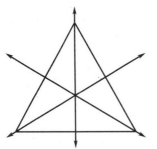

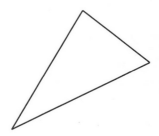

a. Equilateral triangles have three lines of symmetry.

b. Isosceles triangles have one line of symmetry.

c. Scalene triangles do not have any lines of symmetry.

Exercises for Example 2

Find the number of lines of symmetry for the figure described.

9. Rectangle

10. Kite

EXAMPLE 3 *Finding a Minimum Distance*

Find point C on the x-axis so $AC + BC$ is a minimum where A is $(-1, 5)$ and B is $(5, 1)$.

SOLUTION

Reflect A in the x-axis to obtain $A'(-1, -5)$. Then, draw $\overline{A'B}$. Label the point at which this segment intersects the x-axis as C. Because $\overline{A'B}$ represents the shortest distance between A' and B, and $AC = A'C$, you can conclude that at point C a minimum length is obtained. Next, to find the coordinates of C, find an

equation for $\overline{A'B}$. Slope of $\overline{A'B} = \dfrac{1 - (-5)}{5 - (-1)} = \dfrac{6}{6} = 1$

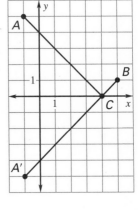

Then use this slope and $A'(-1, -5)$ in $y - y_o = m(x - x_o)$ to get $y + 5 = x + 1$ or $y = x - 4$. Because C is on the x-axis, $y = 0$, so $x = 4$. Therefore, C is $(4, 0)$.

Exercises for Example 3

In Exercises 11–13, find point C on the x-axis so $AC + CB$ is a minimum.

11. $A(-1, -2), B(8, -4)$ **12.** $A(1, 4), B(8, 3)$ **13.** $A(-1.5, 6), B(6, 9)$

Lesson 7.2

NAME _____ DATE _____

Quick Catch-Up for Absent Students

For use with pages 403–410

The items checked below were covered in class on (date missed) _____

Activity 7.2: Reflections in the Plane (p. 403)

_____ **Goal:** Determine the relationship between the line of reflection and the segment connecting a point and its image.

Lesson 7.2: Reflections

_____ **Goal 1:** Identify and use reflections in a plane. (pp. 404–405)

Material Covered:

_____ Example 1: Reflections in a Coordinate Plane

_____ Student Help: Study Tip

_____ Example 2: Proof of Case 1 of Theorem 7.1

_____ Example 3: Finding a Minimum Distance

Vocabulary:

reflection, p. 404   line of reflection, p. 404

_____ **Goal 2:** Identify relationships between reflections and line symmetry. (p. 406)

Material Covered:

_____ Example 4: Finding Lines of Symmetry

_____ Example 5: Identifying Reflections

Vocabulary:

line of symmetry, p. 406

_____ Other (specify) _____

_____

Homework and Additional Learning Support

_____ Textbook (specify) _pp. 407–410_ _____

_____

_____ *Reteaching with Practice* worksheet (specify exercises)_____

_____ *Personal Student Tutor* for Lesson 7.2

NAME _____   DATE _____

Interdisciplinary Application

For use with pages 404–410

Paper Cutting

ART Paper cutting is a folk art that is popular among the common people of China. People often use this art form to express their thoughts, feelings, ideas and dreams. Paper cutting can be more than just a hobby. This art form can be seen on the covers of newspapers and magazines, in business advertisements and as illustrations.

There are five methods of Chinese paper cutting. One is called folded cutting. This technique uses paper that is folded flat and smooth to create geometrical designs that can be used as decorations. The design at the right is an example of the folded paper technique.

In Exercises 1–5, the folded paper technique is used.

1. Your art teacher makes the following design by folding and cutting a piece of paper. Sketch the shape of the cutout that is being reflected. What is the axis of reflection?

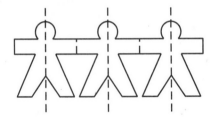

2. You are asked to create a circular figure by first folding a square piece of paper lengthwise, then horizontally, and finally into a triangle as shown at the right. Describe what you think the resulting paper will look like if a cut is made into the folded side of the triangle.

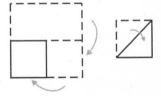

3. Fold and cut a piece of paper as discussed in Exercise 2. Was your answer in Exercise 1 correct? How many lines of symmetry does the figure have? Describe any differences.

4. How many lines of symmetry does the figure at the right have?

5. Using the formula $n(m\angle A) = 180°$, where n is the number of lines of symmetry in the figure, what is the measure each angle in Exercise 4?

Challenge: Skills and Applications

For use with pages 404–410

1. Given a point *P* and a line *k*, explain how a compass and straightedge can be used to construct a reflection of the point in the line.

In Exercises 2 and 3, use a compass and straightedge to construct the reflection of the figure in line *k*.

2.
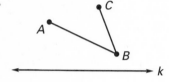

3.
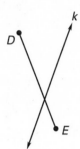

In Exercises 4 and 5, refer to the diagram. Assume that *Y* is the reflection of *X* in line *m*, and *Z* is the reflection of *Y* in *n*.

4. Write a paragraph proof showing that distance *OX* is equal to distance *OZ*.

5. How is ∠*XOZ* related to ∠*SOT*? Write a paragraph proof to defend your answer. (Assume ∠*SOT* is an acute angle with *Y* in its interior, as shown.)

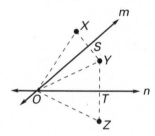

In Exercises 6–15, find the number of lines of symmetry for each type of polygon.

6. equilateral triangle

7. square

8. regular pentagon

9. regular hexagon

10. rhombus (not a square)

11. rectangle (not a square)

12. kite

13. parallelogram (not a rhombus or rectangle)

14. isosceles trapezoid

15. non-isosceles trapezoid

Lesson 7.2

TEACHER'S NAME _____ CLASS _____ ROOM _____ DATE _____

Lesson Plan

2-day lesson (See *Pacing the Chapter*, TE pages 392C–392D)   **For use with pages 411–420**

GOALS
1. **Identify rotations in a plane.**
2. **Use rotational symmetry in real-life situations.**

State/Local Objectives _____

_____

✓ Check the items you wish to use for this lesson.

STARTING OPTIONS
____ Homework Check: TE page 407: Answer Transparencies
____ Warm-Up or Daily Homework Quiz: TE pages 412 and 410, CRB page 39, or Transparencies

TEACHING OPTIONS
____ Motivating the Lesson: TE page 413
____ Lesson Opener (Activity): CRB page 40 or Transparencies
____ Technology Activity with Keystrokes: CRB pages 41–42
____ Examples: Day 1: 1–3, SE pages 413–414; Day 2: 4–5, SE page 415
____ Extra Examples: Day 1: TE pages 413–414 or Transp.; Day 2: TE page 415 or Transp.; Internet
____ Technology Activity: SE page 411
____ Closure Question: TE page 415
____ Guided Practice: SE page 416   Day 1: Exs. 1–9; Day 2: Exs. 10–12

APPLY/HOMEWORK
Homework Assignment
____ Basic   Day 1: 13–30; Day 2: 31–38, 43, 45–54; Quiz 1: 1–8
____ Average   Day 1: 13–30; Day 2: 31–43, 45–54; Quiz 1: 1–8
____ Advanced   Day 1: 13–30; Day 2: 31–54; Quiz 1: 1–8

Reteaching the Lesson
____ Practice Masters: CRB pages 43–45 (Level A, Level B, Level C)
____ Reteaching with Practice: CRB pages 46–47 or Practice Workbook with Examples
____ Personal Student Tutor

Extending the Lesson
____ Applications (Real-Life): CRB page 49
____ Math & History: SE page 420; CRB page 50; Internet
____ Challenge: SE page 419; CRB page 51 or Internet

ASSESSMENT OPTIONS
____ Checkpoint Exercises: Day 1: TE pages 413–414 or Transp.; Day 2: TE page 415 or Transp.
____ Daily Homework Quiz (7.3): TE page 419, CRB page 55, or Transparencies
____ Standardized Test Practice: SE page 419; TE page 419; STP Workbook; Transparencies
____ Quiz (7.1–7.3): SE page 420; CRB page 52

Notes _____

_____

_____

TEACHER'S NAME _____ CLASS _____ ROOM _____ DATE _____

Lesson Plan for Block Scheduling

1-day lesson (See *Pacing the Chapter*, TE pages 392C–392D)   **For use with pages 411–420**

GOALS
1. **Identify rotations in a plane.**
2. **Use rotational symmetry in real-life situations.**

State/Local Objectives _____

_____

_____

✓ **Check the items you wish to use for this lesson.**

STARTING OPTIONS

____ Homework Check: TE page 407: Answer Transparencies
____ Warm-Up or Daily Homework Quiz: TE pages 412 and
     410, CRB page 39, or Transparencies

TEACHING OPTIONS

____ Motivating the Lesson: TE page 413
____ Lesson Opener (Activity): CRB page 40 or Transparencies
____ Technology Activity with Keystrokes: CRB pages 41–42
____ Examples 1–5: SE pages 413–415
____ Extra Examples: TE pages 413–415 or Transparencies; Internet
____ Technology Activity: SE page 411
____ Closure Question: TE page 415
____ Guided Practice Exercises: SE page 416

APPLY/HOMEWORK

Homework Assignment

____ Block Schedule: 13–43, 45–54; Quiz 1: 1–8

Reteaching the Lesson

____ Practice Masters: CRB pages 43–45 (Level A, Level B, Level C)
____ Reteaching with Practice: CRB pages 46–47 or Practice Workbook with Examples
____ Personal Student Tutor

Extending the Lesson

____ Applications (Real-Life): CRB page 49
____ Math & History: SE page 420; CRB page 50; Internet
____ Challenge: SE page 419; CRB page 51 or Internet

ASSESSMENT OPTIONS

____ Checkpoint Exercises: TE pages 413–415 or Transparencies
____ Daily Homework Quiz (7.3): TE page 419, CRB page 55, or Transparencies
____ Standardized Test Practice: SE page 419; TE page 419; STP Workbook; Transparencies
____ Quiz (7.1–7.3): SE page 420; CRB page 52

Notes _____

_____

_____

CHAPTER PACING GUIDE	
Day	**Lesson**
1	7.1 (all)
2	7.2 (all)
3	**7.3 (all)**
4	7.4 (all); 7.5 (begin)
5	7.5 (end); 7.6 (begin)
6	7.6 (end); Review Ch. 7
7	Assess Ch. 7; 8.1 (all)

Geometry
Chapter 7 Resource Book

NAME _____ DATE _____

WARM-UP EXERCISES

For use before Lesson 7.3, pages 411–420

State the definition, theorem, or postulate that justifies each statement.

1. If $\angle ABC \cong \angle A'B'C'$, $\overline{AB} \cong \overline{A'B'}$, and $\overline{BC} \cong \overline{B'C'}$, then $\triangle ABC \cong \triangle A'B'C'$.

2. If $3x + 10 = 15$, then $3x = 5$.

Find the measure of a counterclockwise rotation that would equal each rotation.

3. 180° clockwise rotation

4. 90° clockwise rotation

DAILY HOMEWORK QUIZ

For use after Lesson 7.2, pages 403–410

1. Find the coordinates of $A(3, 2)$ reflected in the line $y = 1$.

2. Find the coordinates of $B(-2, 4)$ reflected in the y-axis.

3. Sketch a hexagon with exactly two lines of symmetry.

4. Given $A(1, -2)$, $B(6, -3)$ find point C on the x-axis so that $AC + BC$ is a minimum.

Lesson 7.3

Activity Lesson Opener

For use with pages 412–420

SET UP: Work in a group.

YOU WILL NEED: • rubber stamps • ink pads • protractor

1. The heart pattern below was made with a single heart-shaped rubber stamp. The preimage was stamped first, then the rubber stamp was rotated 90° counterclockwise and stamped. This rotating and stamping are done three times. With another 90° rotation of the rubber stamp, the pattern repeats. The *angle of rotation* is 90°, and the *center of rotation* is point *C*. Notice that the hearts are equidistant from *C*. How many stamps (hearts) are in this pattern?

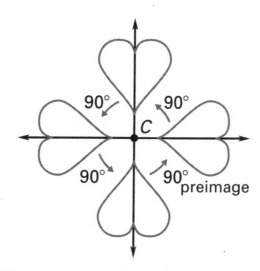

2. Each member of the group should choose an angle of rotation from the chart and use a rubber stamp to make a pattern like the one above. Then share your patterns and complete the table. What happens when you multiply the angle of rotation by the number of stamps in the pattern?

Angle of rotation	Number of stamps in pattern
180°	
120°	
90°	
60°	
45°	

NAME _____ DATE _____

Technology Activity Keystrokes

For use with page 411

TI-92

Construct

1. Draw scalene triangle *ABC*.

   [F3] 3 (Locate desired position for *A*.) [ENTER] *A* (Move to

   location for *B*.) [ENTER] *B* (Move to location for *C*.) [ENTER] *C*

2. Draw intersecting lines *k* and *m* so the lines do not intersect the triangle.

   [F2] 4 (Move cursor to desired location.) [ENTER] (Move cursor to

   draw line *k*.) [ENTER] *k*

   [F2] 4 (Move cursor to desired location.) [ENTER] (Move cursor to

   draw line *m*.) [ENTER] *m*

3. Label the point of intersection of lines *k* and *m* as *P*.

   [F2] 3 (Move cursor to the point of intersection of the two lines.)

   [ENTER] *P*

Investigate

1. Reflect triangle *ABC* in line *k*.

   [F5] 4 (Place cursor on triangle *ABC*.) [ENTER] (Move cursor to

   line *k*.) [ENTER]

   Label the new points on the triangle *A′*, *B′*, and *C′*.

   [F7] 4 (Set cursor on reflected point *A*.) [ENTER] *A* [2nd] [CHAR]

   3 7 [ENTER] [F7] 4 (Set cursor on reflected point *B*.) [ENTER] *B*

   [2nd] [CHAR]

   3 7 [ENTER] [F7] 4 (Set cursor on reflected point *C*.) [ENTER] *C*

   [2nd] [CHAR]

   3 7 [ENTER]

   Reflect triangle *A′B′C′* in line *m*.

   [F5] 4 (Place cursor on triangle *A′B′C′*.) [ENTER] (Move cursor to

   line *m*.) [ENTER]

   Label the new points on the triangle *A″*, *B″*, and *C″*.

   [F7] 4 (Set cursor on reflected point *A*.) [ENTER] *A* [2nd] [11] [ENTER]

   [F7] 4 (Set cursor on reflected point *B*.) [ENTER] *B* [2nd] [11] [ENTER]

   [F7] 4 (Set cursor on reflected point *C*.) [ENTER] *C* [2nd] [11] [ENTER]

3. Draw segments *AP* and *A″P*.

   [F2] 5 (Place cursor on point *A*.) [ENTER] (Move cursor to point *P*.)

   [ENTER]

LESSON
7.3
CONTINUED

NAME _____ DATE _____

Technology Activity Keystrokes

For use with page 411

[F2] 5 (Place cursor on point *A".*) [ENTER] (Move cursor to point *P.*)

[ENTER]

4. Measure angle *APA".*

[F6] 3 (Place cursor on point *A.*) [ENTER] (Move cursor to point *P.*)

[ENTER] (Move cursor to point *P".*) [ENTER]

Measure the acute angle formed by lines *k* and *m.*

[F6] 3 (Place cursor on line *k.*) [ENTER] (Move cursor to point *P.*)

[ENTER] (Move cursor to line *m.*) [ENTER]

5. Find the measures of angles *BPB"* and *CPC"* (to keep the diagram simpler, it is not necessary to draw in the sides of these angles).

[F6] 3 (Place cursor on point *B.*) [ENTER] (Move cursor to point *P.*)

[ENTER] (Move cursor to point *B".*) [ENTER]

[F6] 3 (Place cursor on point *C.*) [ENTER] (Move cursor to point *P.*)

[ENTER] (Move cursor to point *C".*) [ENTER]

SKETCHPAD

Construct

1. Draw a scalene triangle *ABC* using the segment straightedge tool.

2. Draw intersecting lines *k* and *m* using the line straightedge tool.

3. Label the point of intersection of the two lines as *P.* Select point from the toolbox and click on the intersection point. Relabel the point.

Investigate

1. Reflect triangle *ABC* in line *k.* Select line *k.* Choose **Mark Mirror** from the **Transform** menu. Use the selection arrow tool to select the segments and points of triangle *ABC.* Choose **Reflect** from the **Transform** menu. Repeat these steps to reflect triangle *A'B'C'* in line *m.*

3. Draw segments *AP* and *A"P* using the segment straightedge tool. Measure angle *APA".* To measure angle *APA",* use the selection arrow tool to select points *A, P,* and *A".* Then select **Angle** from the **Measure** menu.

4. Measure the acute angle formed by lines *k* and *m.* See Step 4.

5. Find the measures of angles *BPB"* and *CPC"* (to keep the diagram simpler, it is not necessary to draw in the sides of these angles). See Step 5.

NAME _____ DATE _____

Practice A

For use with pages 412–420

Determine whether the figure has rotational symmetry. If so, describe the rotations that map the figure onto itself.

1.
2.
3.
4.

The diagonals of the regular octagon below form eight congruent triangles. Use the diagram to complete the sentence.

5. A clockwise rotation of 45° about *P* maps *A* onto ___?___ .

6. A clockwise rotation of 90° about *P* maps *C* onto ___?___ .

7. A clockwise rotation of 180° about *P* maps ___?___ onto *F*.

8. A counterclockwise rotation of 90° about *P* maps *H* onto ___?___ .

9. A counterclockwise rotation of 45° about *P* maps ___?___ onto *E*.

10. A counterclockwise rotation of 135° about *P* maps ___?___ onto *A*.

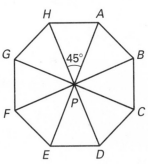

Trace the polygon and point *P* on paper. Then, use a straightedge, compass, and protractor to rotate the polygon clockwise the given number of degrees about *P*.

11. 90°

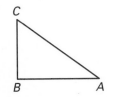

12. 45°

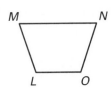

13. 180°

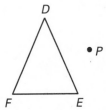

Name the coordinates of the vertices of the image after a clockwise rotation of the given number of degrees about the origin.

14. 90°

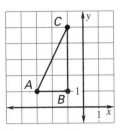

15. 180°

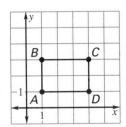

16. 270°

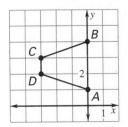

Lesson 7.3

Practice B
For use with pages 412–420

Determine whether the figure has rotational symmetry. If so, describe the rotations that map the figure onto itself.

1.

2.

3.
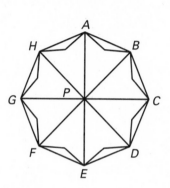

4.

State the segment or triangle that represents the image.

5. 90° clockwise rotation of $\overline{AB}$ about P

6. 90° clockwise rotation of $\overline{DE}$ about P

7. 90° counterclockwise rotation of $\overline{GH}$ about P

8. 180° counterclockwise rotation of $\overline{EF}$ about P

9. 180° clockwise rotation of $\triangle DPE$ about P

10. 45° counterclockwise rotation of $\triangle HPA$ about P

In Exercises 11 and 12, lines ℓ and m intersect at point O. Consider a reflection of $\triangle XYZ$ in ℓ, followed by a reflection in line m.

11. If the angle between ℓ and m is 32°, what is the angle of rotation about O?

12. If the angle of rotation about O is 128°, what is the acute angle between ℓ and m?

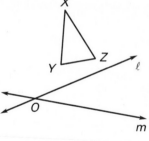

13. Consider two perpendicular lines, ℓ and m. Describe the rotation that is equivalent to reflecting a preimage in ℓ followed by a reflection in m.

Name the coordinates of the vertices of the image after a clockwise rotation of the given number of degrees about the origin.

14. 90°
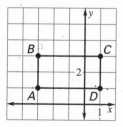

15. 180°
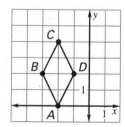

16. 270°
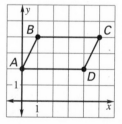

Practice C
For use with pages 412–420

Determine whether the figure has rotational symmetry. If so, describe the rotations that map the figure onto itself.

1.   2.   3.  4.

State the segment or triangle that represents the image.

5. 90° clockwise rotation of $\overline{AB}$ about P

6. 90° clockwise rotation of $\overline{DE}$ about P

7. 90° counterclockwise rotation of $\overline{GH}$ about P

8. 180° counterclockwise rotation of $\overline{EF}$ about P

9. 180° clockwise rotation of $\triangle CJD$ about P

10. 90° counterclockwise rotation of $\triangle GLF$ about P

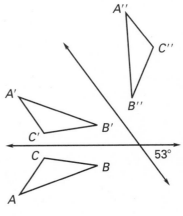

Find the angle of rotation that maps $\triangle ABC$ onto $\triangle A''B''C''$.

11. 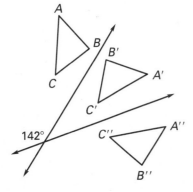 12.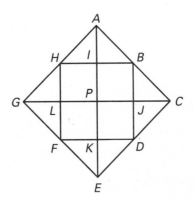

Name the coordinates of the vertices of the image after a clockwise rotation of the given number of degrees about the origin.

13. 90°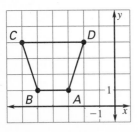

14. 180°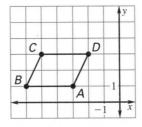

15. 270°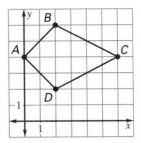

Lesson 7.3

NAME _____ DATE _____

Reteaching with Practice

For use with pages 412–420

GOAL **Identify rotations in a plane.**

VOCABULARY

A **rotation** is a transformation in which a figure is turned about a fixed point.

The fixed point of a rotation is called the **center of rotation.**

Rays drawn from the center of rotation to a point and its image form an angle called the **angle of rotation.**

A figure in the plane has **rotational symmetry** if the figure can be mapped onto itself by a clockwise rotation of 180° or less.

Theorem 7.2 Rotation Theorem
A rotation is an isometry.

Theorem 7.3
If lines *k* and *m* intersect at point *P*, then a reflection in *k* followed by a reflection in *m* is a rotation about point *P*.

The angle of rotation is $2x°$, where $x°$ is the measure of the acute or right angle formed by *k* and *m*.

EXAMPLE 1 *Rotations in a Coordinate Plane*

In a coordinate plane, sketch the quadrilateral whose vertices are
$A(-2, -1)$, $B(-5, 1)$, $C(-4, 5)$, and $D(-1, 2)$. Then, rotate *ABCD* 90°
clockwise about the origin and name the coordinates of the new vertices.
Describe any patterns you see in the coordinates.

SOLUTION

Plot the points. Use a protractor, a compass, and a straightedge to find the rotated vertices. The coordinates of the preimage and image are listed below.

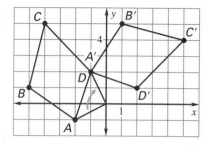

Figure *ABCD*	Figure *A'B'C'D'*
$A(-2, -1)$	$A'(-1, 2)$
$B(-5, 1)$	$B'(1, 5)$
$C(-4, 5)$	$C'(5, 4)$
$D(-1, 2)$	$D'(2, 1)$

In the list above, the *x*-coordinate of the image is the *y*-coordinate of the preimage. The *y*-coordinate of the image is the opposite of the *x*-coordinate of the preimage.

This transformation can be described as $(x, y) \rightarrow (y, -x)$.

Geometry
Chapter 7 Resource Book

NAME _____ DATE _____

Reteaching with Practice

For use with pages 412–420

Exercises for Example 1

In Exercises 1 and 2, use the given information to rotate the quadrilateral. Name the vertices of the image and compare with the vertices of the preimage. Describe any patterns you see.

1. 90° clockwise about origin

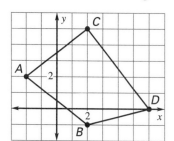

2. 180° counterclockwise about origin

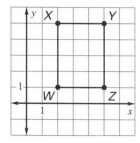

 **EXAMPLE 2** *Identifying Rotational Symmetry*

Which figures have rotational symmetry? For those that do, describe the rotations that map the figure onto itself.

a. Isosceles triangle **b.** Kite **c.** Rhombus

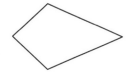

SOLUTION

a. The isosceles triangle does not have rotational symmetry.

b. This kite has rotational symmetry. It can be mapped onto itself by a rotation of 180° about its center.

c. This rhombus has rotational symmetry. It can be mapped onto itself by a rotation of 180° about its center.

Exercises for Example 2

Decide which figures have rotational symmetry. For those that do, describe the rotations that map the figure onto itself.

3. Equilateral triangle **4.** Rectangle **5.** Regular pentagon

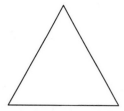

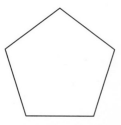

Lesson 7.3

Quick Catch-Up for Absent Students

For use with pages 411–420

The items checked below were covered in class on (date missed) _____

Activity 7.3: Investigating Double Reflections

_____ **Goal:** Use geometry software to discover the type of transformation that results when a triangle is reflected twice in the plane. (p. 411)

_____ Student Help: Software Help

Lesson 7.3: Rotations

_____ **Goal 1:** Identify rotations in a plane. (pp. 412–414)

Material Covered:

_____ Example 1: Proof of Theorem 7.2

_____ Student Help: Look Back

_____ Activity: Rotating a Figure

_____ Example 2: Rotations in a Coordinate Plane

_____ Example 3: Using Theorem 7.3

Vocabulary:

rotation, p. 412   center of rotation, p. 412
angle of rotation, p. 412

_____ **Goal 2:** Use rotational symmetry in real-life situations. (p. 415)

Material Covered:

_____ Example 4: Identifying Rotational Symmetry

_____ Example 5: Using Rotational Symmetry

Vocabulary:

rotational symmetry, p. 415

_____ Other (specify) _____

_____

Homework and Additional Learning Support

_____ Textbook (specify) _pp. 416–420_____

_____

_____ Internet: Extra Examples at www.mcdougallittell.com

_____ *Reteaching with Practice* worksheet (specify exercises)_____

_____ *Personal Student Tutor* for Lesson 7.3

NAME _____ DATE _____

Real-Life Application: When Will I Ever Use This?

For use with pages 412–420

Ferris Wheel

A Ferris wheel is an amusement ride consisting of a power operated vertical wheel that is about 50 feet in diameter. It has a steel frame and along the rim of the Ferris wheel there are seats that are suspended freely so that they remain upright as they revolve. The Ferris wheel is not a "thrill-seeking" ride; rather, it is a ride that allows its passengers to see sights around them from a high vantage point.

The first Ferris wheel was built by George Washington Ferris in 1893 for the World's Fair in Chicago. This wheel was impressive with its 250 foot diameter and 36 cars that held 40 passengers each. Today, the largest Ferris wheel is called "The London Eye." It is located on the south bank of the Thames River. This Ferris Wheel is 443 feet tall and has 32 glass-enclosed capsules that hold 25 people. One rotation of the wheel takes 30 minutes, allowing each passenger to enjoy a spectacular view of London.

In Exercises 1–5, use the diagram at the right of a Ferris wheel.

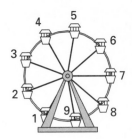

1. Find the measure of the angle between any two seats.

2. You are at seat position 1. At what position will you be after a 120° clockwise rotation about the center?

3. You are at seat position 7. At what position will you be after an 80° counterclockwise rotation about the center?

4. Suppose the Ferris wheel takes 9 minutes to make one complete revolution. How long would it take you to move from seat position 3 to seat position 8?

NAME _____ DATE _____

Math and History Application

For use with page 420

For thousands of years, cultures have constructed buildings with symmetric patterns. These patterns add interest and give a building character.

Many people consider the most beautiful building in the world to be the Taj Mahal, located in India. Built between 1631 and 1653 by the emperor Shah Jahan as a monument to his wife, Mumtaz Mahal, the Taj Mahal provides many stunning examples of visual symmetry.

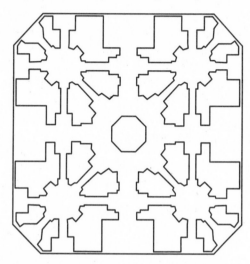

1. Consider the floor map of the Taj Mahal shown above. How many lines of symmetry does the floor map have?

2. Does the floor map have rotational symmetry? If so, describe a rotation that maps the pattern onto itself.

Challenge: Skills and Applications

For use with pages 412–420

1. Suppose you are given points P, X, X', and S such that X' is the image of X under a rotation about P. Explain how a compass and straightedge can be used (without a protractor) to construct the image of S under the same rotation.

In Exercises 2–3, X' is the image of X under a rotation about P. Use a compass and straightedge to construct the image of S under the same rotation.

2.

3.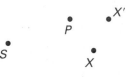

In Exercises 4–7, refer to the diagram, where Q' and R' are the images of Q and R after a 90° counterclockwise rotation about P.

4. Find the coordinates of Q' in terms of a, b, x_0, and y_0.

5. Find the coordinates of R' in terms of a, b, x_0, and y_0.

6. If the point $(5, 3)$ is rotated 90° counterclockwise about $(0, 0)$, what are the coordinates of the image point?

7. If the point $(2, -5)$ is rotated 90° counterclockwise about $(-3, 7)$, what are the coordinates of the image point?

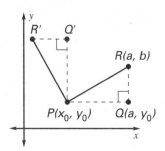

In Exercises 8–15, determine whether the polygon has rotational symmetry. If it does, describe the rotations that map the polygon onto itself.

8. equilateral triangle

9. square

10. regular pentagon

11. regular hexagon

12. rhombus (not a square)

13. rectangle (not a square)

14. kite

15. isosceles trapezoid

Lesson 7.3

NAME _____ DATE _____

Quiz 1

For use after Lessons 7.1–7.3

Use the transformation at the right. *(Lesson 7.1)*

1. Figure $ABCD \rightarrow$ Figure ___?___.

2. Name and describe the transformation.

3. Is the transformation an isometry? Explain.

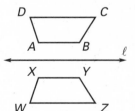

Answers

1. _____

2. _____

3. _____

   _____

4. _____

5. _____

6. _____

7. _____

8. _____

   _____

   _____

In Exercises 4–7, find the coordinates of the reflection without using a coordinate plane. *(Lesson 7.2)*

4. $A(1, 3)$ reflected in the *x*-axis

5. $B(-2, -3)$ reflected in the *y*-axis

6. $C(-2, 0)$ reflected in the *x*-axis

7. $D(5.2, -2)$ reflected in the *y*-axis

8. Use Figure 1 below to describe the transformation needed to create Figure 2. *(Lesson 7.3)*

Figure 1

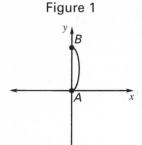

Figure 2

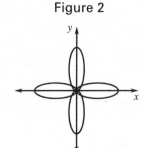

TEACHER'S NAME _____ CLASS _____ ROOM _____ DATE _____

Lesson Plan

1-day lesson (See *Pacing the Chapter,* TE pages 392C–392D)   For use with pages 421–428

GOALS
1. **Identify and use translations in the plane.**
2. **Use vectors in real-life situations.**

State/Local Objectives _____

_____

✓ Check the items you wish to use for this lesson.

STARTING OPTIONS
____ Homework Check: TE page 416: Answer Transparencies
____ Warm-Up or Daily Homework Quiz: TE pages 421 and 419, CRB page 55, or Transparencies

TEACHING OPTIONS
____ Lesson Opener (Application): CRB page 56 or Transparencies
____ Examples 1–6: SE pages 422–424
____ Extra Examples: TE pages 422–424 or Transparencies; Internet
____ Closure Question: TE page 424
____ Guided Practice Exercises: SE page 425

APPLY/HOMEWORK
Homework Assignment
____ Basic   16–46 even, 56–59, 62–72 even
____ Average   16–46 even, 43, 53–59, 62–72 even
____ Advanced   16–46 even, 43, 53–61, 62–72 even

Reteaching the Lesson
____ Practice Masters: CRB pages 57–59 (Level A, Level B, Level C)
____ Reteaching with Practice: CRB pages 60–61 or Practice Workbook with Examples
____ Personal Student Tutor

Extending the Lesson
____ Cooperative Learning Activity: CRB page 63
____ Applications (Interdisciplinary): CRB page 64
____ Challenge: SE page 428; CRB page 65 or Internet

ASSESSMENT OPTIONS
____ Checkpoint Exercises: TE pages 422–424 or Transparencies
____ Daily Homework Quiz (7.4): TE page 428, CRB page 68, or Transparencies
____ Standardized Test Practice: SE page 428; TE page 428; STP Workbook; Transparencies

Notes _____

_____

_____

LESSON 7.4

Lesson Plan for Block Scheduling

Half-day lesson (See *Pacing the Chapter,* TE pages 392C–392D)   For use with pages 421–428

 GOALS
1. **Identify and use translations in the plane.**
2. **Use vectors in real-life situations.**

State/Local Objectives _____

_____

_____

CHAPTER PACING GUIDE	
Day	Lesson
1	7.1 (all)
2	7.2 (all)
3	7.3 (all)
4	**7.4 (all)**; 7.5 (begin)
5	7.5 (end); 7.6 (begin)
6	7.6 (end); Review Ch. 7
7	Assess Ch. 7; 8.1 (all)

✓ **Check the items you wish to use for this lesson.**

STARTING OPTIONS
____ Homework Check: TE page 416: Answer Transparencies
____ Warm-Up or Daily Homework Quiz: TE pages 421 and
        419, CRB page 55, or Transparencies

TEACHING OPTIONS
____ Lesson Opener (Application): CRB page 56 or Transparencies
____ Examples 1–6: SE pages 422–424
____ Extra Examples: TE pages 422–424 or Transparencies; Internet
____ Closure Question: TE page 424
____ Guided Practice Exercises: SE page 425

APPLY/HOMEWORK
Homework Assignment (See also the assignment for Lesson 7.5.)
____ Block Schedule: 16–46 even, 43, 53–59, 62–72 even

Reteaching the Lesson
____ Practice Masters: CRB pages 57–59 (Level A, Level B, Level C)
____ Reteaching with Practice: CRB pages 60–61 or Practice Workbook with Examples
____ Personal Student Tutor

Extending the Lesson
____ Cooperative Learning Activity: CRB page 63
____ Applications (Interdisciplinary): CRB page 64
____ Challenge: SE page 428; CRB page 65 or Internet

ASSESSMENT OPTIONS
____ Checkpoint Exercises: TE pages 422–424 or Transparencies
____ Daily Homework Quiz (7.4): TE page 428, CRB page 68, or Transparencies
____ Standardized Test Practice: SE page 428; TE page 428; STP Workbook; Transparencies

Notes _____

_____

_____

Lesson 7.4

Geometry
Chapter 7 Resource Book

WARM-UP EXERCISES

For use before Lesson 7.4, pages 421–428

The vertices of △KLM are K(3, 3), L(5, 4), and M(0, 6). Find the new vertices after each slide.

1. left two units   **2.** down five units

The coordinates of the endpoints of $\overline{PQ}$ are P(1, 2) and Q(4, 6).

3. What is the length of $\overline{PQ}$?

4. If $\overline{PQ}$ is translated right five units, what is the length of its image?

···

DAILY HOMEWORK QUIZ

For use after Lesson 7.3, pages 411–420

1. Draw a square. Below the square draw a point P. Then use a straightedge, compass, and protractor to rotate the square clockwise 60° about P.

2. In a coordinate plane, sketch the triangle whose vertices are $A(2, -2)$, $B(4, 1)$, $C(5, 1)$. Then rotate △ABC 90° clockwise about $(0, 0)$ and name the coordinates of the new vertices.

3. Lines m and n intersect at D. If △ABC is reflected in line m followed by a reflection in line n, what is the angle of rotation about D, when the measure of the acute angle between m and n is 25°?

NAME _____ DATE _____

Application Lesson Opener

For use with pages 421–428

Large airports have many different ways of transporting people and their baggage. In the story below, Abigail is transported by three different mechanical devices. Each time, Abigail is moved from one point to another in a straight line, which models a *translation*.

> Abigail arrives at the airport, parks her car in the parking garage, and walks to an elevator. She gets in the elevator at the fourth floor and goes to the first floor to check in. After checking in, she takes an escalator to the second floor and walks to the gate area. Instead of walking from Gate 1 to Gate 17, she stands on the moving walkway.

1. Name the three mechanical devices in the story that transport Abigail. Match each device with a figure below that shows her movement. Point *A* represents Abigail, and the image point *A'* represents her new position after the translation.

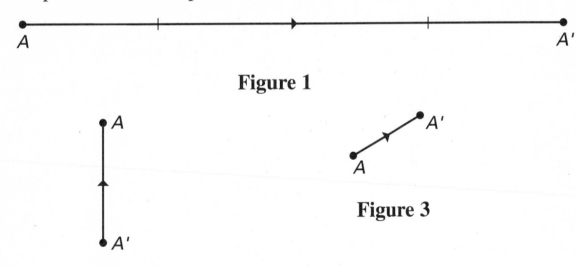

Figure 1

Figure 2

Figure 3

2. For each of the three mechanical devices, write a sentence that describes the change in position of Abigail. Your descriptions should include answers to some of the following questions. Was her movement *horizontal*, *vertical*, or *both*? What was the direction of her movement? How far did she move?

NAME _____ DATE _____

Practice A
For use with pages 421–428

Four playing cards are face down on the floor. Name the transformation.

A    B

C    D

1. Shape *A* is mapped to shape *C*.

3. Shape *C* is mapped to shape *D*.

2. Shape *A* is mapped to shape *B*.

4. Shape *B* is mapped to shape *C*.

Match the graph with the description of the translation.

A.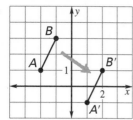

B.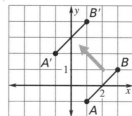

C.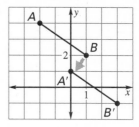

5. $(x, y) \rightarrow (x + 2, y - 3)$

6. $(x, y) \rightarrow (x + 3, y - 2)$

7. $(x, y) \rightarrow (x - 2, y + 3)$

Consider the translation that is defined by the coordinate notation
$(x, y) \rightarrow (x + 4, y - 6)$.

8. What is the image of $(1, 1)$?

10. What is the image of $(5, 2)$?

12. What is the image of $(3, 0)$?

9. What is the image of $(-3, 4)$?

11. What is the preimage of $(0, 4)$?

13. What is the preimage of $(2, -4)$?

Copy figure *ABCD* and draw its image after the translation.

14. $(x, y) \rightarrow (x + 1, y + 4)$

15. $(x, y) \rightarrow (x - 2, y - 2)$

16. $(x, y) \rightarrow (x - 5, y + 6)$

17. $(x, y) \rightarrow (x + 3, y - 5)$

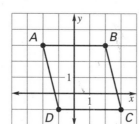

NAME _____ DATE _____

Practice B

For use with pages 421–428

Describe the translation using (a) coordinate notation and (b) a vector in component form.

1.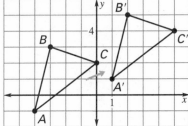

2.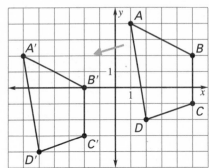

Name the vector and write its component form.

3.

4.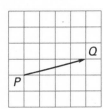

5.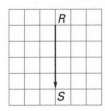

Consider the translation that is defined by the coordinate notation
$(x, y) \rightarrow (x - 5, y + 8)$.

6. What is the image of $(4, 2)$?

7. What is the image of $(-1, 5)$?

8. What is the preimage of $(-3, -4)$?

9. What is the preimage of $(7, -5)$?

10. What is the image of $(0, 2)$?

11. What is the preimage of $(-4, 6)$?

Use a straightedge and graph paper to translate △*ABC* by the given vector.

12. $\langle -3, 1 \rangle$

13. $\langle 2, -3 \rangle$

14. $\langle 4, -1 \rangle$

15. $\langle -5, -2 \rangle$

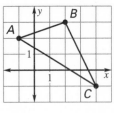

Use the figure at the right which shows the distance between lines ℓ and *m* to be 4.

16. What is the length $\overline{BB''}$?

17. What is the length of $\overline{AA''}$?

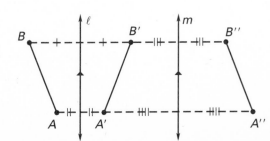

NAME _____ DATE _____

Practice C

For use with pages 421–428

Use coordinate notation to describe the translation.

1. 4 units to the right and 3 units up

2. 5 units left and 2 units down

3. 1 unit to the left and 1 unit up

4. 3 units down

5. 7 units to the left and 4 units down

6. 10 units right and 8 units up

Name the vector and write its component form.

7.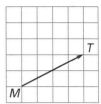

8.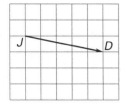

9.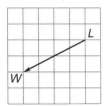

Given △ABC with vertices A(−2, 4), B(6, 2), and C(3, −2) is translated to △A′B′C′. Determine the translation using a vector in component from, and determine the coordinates of the remaining vertices.

10. $A'(3, -2)$

11. $C'(3, 4)$

12. $A'(-5, 5)$

13. $B'(2, -5)$

14. $C'(-4, -5)$

15. $B'(8, 6)$

Copy quadrilateral *MNOP* and draw its image after the translation.

16. $(x, y) \rightarrow (x - 2, y + 4)$

17. $(x, y) \rightarrow (x + 5, y + 1)$

18. $(x, y) \rightarrow (x - 3, y - 7)$

19. $(x, y) \rightarrow (x + 4, y - 5)$

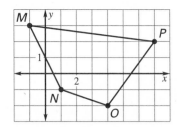

20. Write a paragraph proof for a portion of Theorem 7.5.

   Given: $k \parallel m$

        P' is the reflection of P in line k.

        P'' is the reflection of P' in line m.

   Prove: $PP'' = 2d$, where d is the distance between k and m.

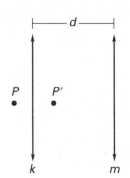

NAME _____ DATE _____

Reteaching with Practice

For use with pages 421–428

GOAL **Identify and use translations in the plane.**

VOCABULARY

A **translation** is a transformation that maps every two points P and Q in the plane to points P' and Q', so that the following properties are true: 1) $PP' = QQ'$ and 2) $\overline{PP'} \parallel \overline{QQ'}$, or $\overline{PP'}$ and $\overline{QQ'}$ are collinear.

A **vector** is a quantity that has both direction and *magnitude*, or size.

When a vector is drawn as ray $\overrightarrow{PQ}$, the **initial point,** or starting point, of the vector is point P and the **terminal point,** or ending point, of the vector is point Q.

The **component form** of a vector combines the horizontal and vertical components.

EXAMPLE 1 *Using Theorem 7.5*

In the diagram, a reflection in line k maps $\overline{AB}$ to $\overline{A'B'}$, a reflection in line m maps $\overline{A'B'}$ to $\overline{A''B''}$, $k \parallel m$, $AW = 7$, and $ZA'' = 3$.

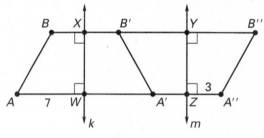

a. Name some congruent segments.

b. Does $WZ = XY$? Explain.

c. What is the length of $\overline{AA''}$?

SOLUTION

a. Here are some sets of congruent segments: $\overline{AB}$, $\overline{A'B'}$, and $\overline{A''B''}$; $\overline{BX}$ and $\overline{XB'}$; $\overline{B'Y}$ and $\overline{YB''}$.

b. Yes, $WZ = XY$ because $\overline{WZ}$ and $\overline{XY}$ are opposite sides of a rectangle.

c. Because $AA'' = BB''$, the length of $\overline{AA''}$ is $7 + 7 + 3 + 3$, or 20 units.

Exercises for Example 1

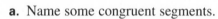

In the diagram $k \parallel m$, $\triangle XYZ$ is reflected in line k, and $\triangle X'Y'Z'$ is reflected in line m.

1. Name two segments parallel to $\overline{YY''}$.

2. If the length of $\overline{ZZ''}$ is 6 cm, what is the distance between k and m?

3. A translation maps $\triangle XYZ$ onto which triangle?

4. Which lines are perpendicular to $\overline{XX''}$?

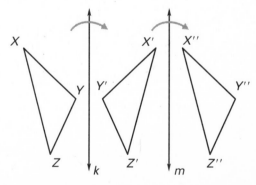

Lesson 7.4

NAME _____ DATE _____

Reteaching with Practice

For use with pages 421–428

EXAMPLE 2 *Translations in a Coordinate Plane*

Sketch a quadrilateral with vertices $A(0, 4)$, $B(-2, 1)$, $C(0, -3)$, and $D(3, 4)$. Then sketch the image of the quadrilateral after the translation $(x, y) \rightarrow (x + 2, y - 1)$.

SOLUTION

Plot the points as shown. Shift each point 2 units to the right and 1 unit down to find the translated vertices.

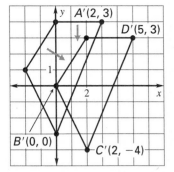

Exercises for Example 2

In Exercises 5–8, copy figure *PQRS* and draw its image after the translation.

   5. $(x, y) \rightarrow (x - 4, y + 1)$

   6. $(x, y) \rightarrow (x, y - 5)$

   7. $(x, y) \rightarrow (x - 2, y - 2)$

   8. $(x, y) \rightarrow (x + 7, y + 3)$

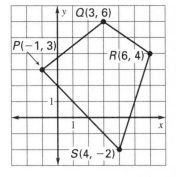

EXAMPLE 3 *Finding Vectors*

In the diagram, $\triangle ABC$ maps onto $\triangle A'B'C'$ by a translation. Write the component form of the vector that can be used to describe the translation.

SOLUTION

Choose any vertex and its image, say A and A'. To move from A to A', you move 3 units to the right and 5 units down. The component form of the vector is $\langle 3, -5 \rangle$.

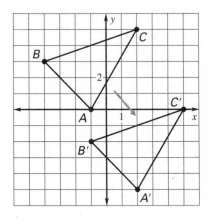

Exercises for Example 3

In Exercises 9 and 10, write the component form of the vector that describes the translation which maps $\triangle ABC$ onto $\triangle A'B'C'$.

   9. $A(3, 6)$, $B(1, 0)$, $C(4, 8)$; $A'(1, 2)$, $B'(-1, -4)$, $C'(2, 4)$

  10. $A(-6, -2)$, $B(-5, 3)$, $C(1, -1)$; $A'(-3, -5)$, $B'(-2, 0)$, $C'(4, -4)$

NAME _____ DATE _____

Quick Catch-Up for Absent Students

For use with pages 421–428

The items checked below were covered in class on (date missed) _____

Lesson 7.4: Translations and Vectors

_____ **Goal 1:** Identify and use translations in the plane. (pp. 421–422)

Material Covered:

_____ Student Help: Study Tip

_____ Example 1: Using Theorem 7.5

_____ Example 2: Translations in a Coordinate Plane

Vocabulary:

translation, p. 421

_____ **Goal 2:** Use vectors in real-life situations. (pp. 423–424)

Material Covered:

_____ Student Help: Study Tip

_____ Example 3: Identifying Vector Components

_____ Example 4: Translation Using Vectors

_____ Example 5: Finding Vectors

_____ Example 6: Using Vectors

Vocabulary:

vector, p. 423   initial point, p. 423

terminal point, p. 423   component form, p. 423

_____ Other (specify) _____

_____

Homework and Additional Learning Support

_____ Textbook (specify) _pp. 425–428_____

_____

_____ Internet: Extra Examples at www.mcdougallittell.com

_____ *Reteaching with Practice* worksheet (specify exercises)_____

_____ *Personal Student Tutor* for Lesson 7.4

NAME _____ DATE _____

Cooperative Learning Activity

For use with pages 421–428

GOAL **To investigate vectors in moving a heavy object**

Materials: length of string measuring approximately one yard

Exploring Vectors

A vector is a quantity with both magnitude and direction. One example of a vector is force. A force can be thought of as a push or a pull. By using the components of a vector, it is possible to use a relatively small force to move a relatively large object.

Instructions

1 Tie one end of the string to a fixed, sturdy object. For example, use the leg of a lab table that is attached to the floor (or the hinge of a door or closet).

2 Tie the other end of the string to an object that weighs about 15–25 pounds (the leg of a mobile student desk works well). Separate the two objects so that the string is taut.

3 Hold the string in the center and pull towards the fixed object. Move the object approximately 8 inches. Have each member of the group perform this activity.

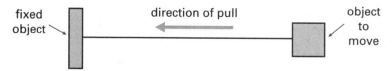

4 Return the objects to a position so that the string is taut once again.

5 Pull the string in the direction shown below and move the object approximately 8 inches.

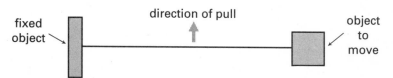

Analyzing the Results

1. How does the amount of force needed pulling in Step 3 compare to the force when pulling in Step 5?

2. How do vectors play a role in moving the object?

3. Would this work with a much heavier object, such as a car?

Interdisciplinary Application

For use with pages 421–428

Avalanches

GEOLOGY   An avalanche is defined as a rapidly descending mass of snow, ice, rock or a mixture of these materials down a mountainous slope. Disturbances such as earthquakes, high winds, or explosions can trigger an avalanche. Avalanches can attain speeds over 200 miles per hour and can completely wipe out anything in its path. One of the most devastating avalanches on record occurred on Mount Huascaran, the highest mountain in Peru, in 1962. Approximately 6000 people were killed when a hanging glacier broke off from a part of the mountain and swept down its slope burying villages in its path.

In order to find people who have been buried by avalanches, rescue teams some-times use trained dogs. The dog's keen sense of smell can help locate survivors and once the dog locates a point, the team as well as the dog begins the strenu-ous task of digging through the snow to reach the survivor.

In Exercises 1–4, use the following information.

Suppose a skier is buried by an avalanche and the skier's position is represented by point $D(12, 2)$ in the coordinate plane. A rescue dog picks up the skier's scent at point $A(0, 16)$ and begins digging. The team reaches point $B(6, 9)$ and then digs off course to point $C(7, 4)$ as shown below.

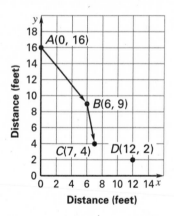

1. How far away is the skier when the dog begins to dig?

2. Write the component forms of the two vectors in the diagram.

3. Write the component form of the vector that describes the path the rescue team can take to arrive at $D(12, 2)$.

4. Suppose the rescue team dug directly to the trapped skier without digging off course. Write the component form of the vector that describes the rescue.

Geometry
Chapter 7 Resource Book

NAME _____ DATE _____

Challenge: Skills and Applications

For use with pages 421–428

1. Suppose you are given points X, X', and Y that are not collinear. If X' is the image of X under a translation, explain how a compass and straightedge can be used to construct the image of Y under the same translation.
   (Hint: What kind of polygon is $XX'Y'Y$?)

In Exercises 2 and 3, X' is the image of X under a translation. Use a compass and straightedge to construct the image of $\overline{YZ}$ under the same translation.

2.

3.

In Exercises 4 and 5, let P' and Q' be the images of P and Q, respectively, under an isometry. Determine whether the statement is true or false. If it is true, give a proof using coordinate geometry. If it is false, give a counterexample.

4. If the isometry is the translation $(x, y) \rightarrow (x + h, y + k)$, where h and k are constants, then $\overline{PP'}$ and $\overline{QQ'}$ are congruent and parallel (or collinear).

5. If $\overline{PP'}$ and $\overline{QQ'}$ are congruent and parallel, then the isometry is a translation.

6. Jody used the file *wave.gif* to create the background for a website. When the page is viewed, *wave.gif* is tiled as shown.

   Assume that *wave.gif* has a width of 90 units and a height of 50 units.

   If $(x, y) \rightarrow (x + h, y + k)$ maps a point on the background to a corresponding point on the background, what can you say about h and k?

wave.gif →

TEACHER'S NAME _____ CLASS _____ ROOM _____ DATE _____

Lesson Plan

2-day lesson (See *Pacing the Chapter,* TE pages 392C–392D)   For use with pages 429–436

GOALS
1. **Identify glide reflections in a plane.**
2. **Represent transformations as compositions of simpler transformations.**

State/Local Objectives _____

_____

✓ Check the items you wish to use for this lesson.

STARTING OPTIONS

____ Homework Check: TE page 425: Answer Transparencies
____ Warm-Up or Daily Homework Quiz: TE pages 430 and 428, CRB page 68, or Transparencies

TEACHING OPTIONS

____ Concept Activity: SE page 429
____ Lesson Opener (Application): CRB page 69 or Transparencies
____ Technology Activity with Keystrokes: CRB pages 70–73
____ Examples: Day 1: 1–3, SE pages 430–431; Day 2: 4–5, SE page 432
____ Extra Examples: Day 1: TE page 431 or Transp.; Day 2: TE page 432 or Transp.
____ Closure Question: TE page 432
____ Guided Practice: SE page 433   Day 1: Exs. 1–7; Day 2: Ex. 8

APPLY/HOMEWORK
Homework Assignment

____ Basic   Day 1: 9–21; Day 2: 22–27, 29, 30, 32–34, 39, 41–45, 46–54 even
____ Average   Day 1: 9–21; Day 2: 22–27, 29, 30, 32–39, 41–45, 46–54 even
____ Advanced   Day 1: 9–21; Day 2: 22–27, 29, 30, 32–45, 46–54 even

Reteaching the Lesson

____ Practice Masters: CRB pages 74–76 (Level A, Level B, Level C)
____ Reteaching with Practice: CRB pages 77–78 or Practice Workbook with Examples
____ Personal Student Tutor

Extending the Lesson

____ Applications (Real-Life): CRB page 80
____ Challenge: SE page 436; CRB page 81 or Internet

ASSESSMENT OPTIONS

____ Checkpoint Exercises: Day 1: TE page 431 or Transp.; Day 2: TE page 432 or Transp.
____ Daily Homework Quiz (7.5): TE page 436, CRB page 84, or Transparencies
____ Standardized Test Practice: SE page 436; TE page 436; STP Workbook; Transparencies

Notes _____

_____

_____

TEACHER'S NAME _____ CLASS _____ ROOM _____ DATE _____

Lesson Plan for Block Scheduling

1-day lesson (See *Pacing the Chapter,* TE pages 392C–392D)   **For use with pages 429–436**

GOALS 1. **Identify glide reflections in a plane.**
2. **Represent transformations as compositions of simpler transformations.**

State/Local Objectives _____

_____

_____

CHAPTER PACING GUIDE	
Day	**Lesson**
1	7.1 (all)
2	7.2 (all)
3	7.3 (all)
4	7.4 (all); **7.5 (begin)**
5	**7.5 (end)**; 7.6 (begin)
6	7.6 (end); Review Ch. 7
7	Assess Ch. 7; 8.1 (all)

✓ **Check the items you wish to use for this lesson.**

STARTING OPTIONS
____ Homework Check: TE page 425: Answer Transparencies
____ Warm-Up or Daily Homework Quiz: TE pages 430 and
        428, CRB page 68, or Transparencies

TEACHING OPTIONS
____ Concept Activity: SE page 429
____ Lesson Opener (Application): CRB page 69 or Transparencies
____ Technology Activity with Keystrokes: CRB pages 70–73
____ Examples: Day 4: 1–3, SE pages 430–431; Day 5: 4–5, SE page 432
____ Extra Examples: Day 4: TE page 431 or Transp.; Day 5: TE page 432 or Transp.
____ Closure Question: TE page 432
____ Guided Practice: SE page 433   Day 4: Exs. 1–7; Day 5: Ex. 8

APPLY/HOMEWORK
Homework Assignment (See also the assignments for Lessons 7.4 and 7.6.)
____ Block Schedule: Day 4: 9–21; Day 5: 22–27, 29, 30, 32–39, 41–45, 46–54 even

Reteaching the Lesson
____ Practice Masters: CRB pages 74–76 (Level A, Level B, Level C)
____ Reteaching with Practice: CRB pages 77–78 or Practice Workbook with Examples
____ Personal Student Tutor

Extending the Lesson
____ Applications (Real-Life): CRB page 80
____ Challenge: SE page 436; CRB page 81 or Internet

ASSESSMENT OPTIONS
____ Checkpoint Exercises: Day 4: TE page 431 or Transp.; Day 5: TE page 432 or Transp.
____ Daily Homework Quiz (7.5): TE page 436, CRB page 84, or Transparencies
____ Standardized Test Practice: SE page 436; TE page 436; STP Workbook; Transparencies

Notes _____

_____

_____

NAME _____ DATE _____

WARM-UP EXERCISES

For use before Lesson 7.5, pages 429–436

$\overline{PQ}$ has endpoints $P(-4, -4)$ and $Q(-1, -3)$. Find the coordinates of P' and Q' after each translation.

1. $(x, y) \rightarrow (x, y + 3)$

2. $(x, y) \rightarrow (x + 1, y - 1)$

3. Find the coordinates of the endpoints of $\overline{P'Q'}$ after $\overline{PQ}$ is rotated 180° about the origin.

4. Find the coordinates of $\overline{P'Q'}$ after $\overline{PQ}$ is reflected in the x-axis.

· ·

DAILY HOMEWORK QUIZ

For use after Lesson 7.4, pages 421–428

1. Describe the translation using **(a)** coordinate notation and **(b)** a vector in component form.

2. Name the vector and write its component form.

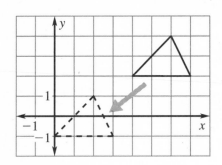

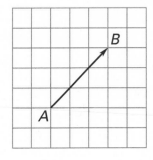

3. Consider the translation that is defined by the coordinate notation $(x, y) \rightarrow (x + 4, y - 1)$.

a. What is the image of $(2, 5)$?

b. What is the preimage of $(-1, 3)$?

4. The vertices of $\triangle ABC$ are $A(-5, 3)$, $B(4, 2)$, and $C(-1, -1)$. Name the vector that describes a translation such that $A'(-2, -1)$, $B'(7, -2)$, and $C'(2, -5)$.

NAME _____ DATE _____

Application Lesson Opener

For use with pages 430–436

Lesson 7.5

YOU WILL NEED: • tracing paper • ruler

Wallpaper patterns usually extend both horizontally and vertically to cover a wall with any dimensions. The wallpaper pattern shown at the right is based on only one figure, the flower. Two types of transformations, *translations* and *reflections*, are applied to the flower to extend the pattern.

1. Trace the circled flower and see how applying translations up and down gives a vertical column of flowers as shown in the pattern.

2. Trace the circled flower and see how applying translations right and left gives a horizontal row of flowers as shown in the pattern.

3. Trace the circled flower and apply the following two transformations: (1) Reflect the flower in a vertical line halfway between the flower and the next column of flowers. (2) Translate this image up, half the distance the flower in Exercise 1 was translated. Describe the final image.

4. Now trace the original circled flower again and reverse the order of the two transformations given in Exercise 3. Translate the flower up, then reflect this image in the vertical line. Compare the final image with the one from Exercise 3. What do you notice?

5. Explain how the transformations in Exercises 3 and 4 were alike and how they were different.

NAME _____ DATE _____

Technology Activity

For use with pages 430–436

GOAL **To use geometry software to find the image of a composition**

When two or more transformations are combined to produce a single transformation, the result is called a *composition* of the transformation. The order the transformations are performed may or may not affect the final image. You can use geometry software to perform the order of the transformations as they are given. You can then perform the transformations in reverse order and verify the affect (if any) on the final image. Consider the image of $\overline{AB}$ after a composition of the given rotation and reflection.

$A(1, -0.5)$, $B(1.5, -1.5)$

Rotation: 90° counterclockwise about the origin

Reflection: in the y-axis

Activity

❶ Draw $\overline{AB}$. (Use the grid and axes features in rectangular mode.)

❷ Rotate $\overline{AB}$ 90° counterclockwise about the origin.

❸ Use the software's reflection feature to reflect the result of Step 2 in the y-axis.

❹ Perform the transformations in reverse order. First, use the software's reflection feature to reflect $\overline{AB}$ in the y-axis.

❺ Rotate the result of Step 4 90° counterclockwise about the origin. Did the order of the transformations affect the final image?

Exercises

Use geometry software to draw the image of $\overline{AB}$ after a composition using the given transformations in the order they appear. Then, perform the transformations in reverse order. Does the order affect the final image?

1. $A(1, -0,5)$, $B(1.5, -1.5)$
   Rotation: 180° counterclockwise about the origin
   Reflection: in the y-axis

2. $A(1, 1)$, $B(1, -1)$
   Rotation: 90° counterclockwise about the origin
   Reflection: in the y-axis

3. $A(-1, 1)$, $B(1, 1)$
   Rotation: 90° counterclockwise about the origin
   Reflection: in the y-axis

Geometry
Chapter 7 Resource Book

Technology Activity Keystrokes

For use with pages 430–436

TI-92

1. Draw $\overline{AB}$.

F8 9 (Set Coordinate Axes to RECTANGULAR, Grid to ON, and Angle to DEGREE.) **ENTER**

F2 5 (Move cursor to point $(1, -0.5)$ and prompt says "ON THIS POINT OF THE GRID".) **ENTER** A (Move cursor to point $(1.5, -1.5)$ and prompt says "ON THIS POINT OF THE GRID".) **ENTER** B

2. Rotate $\overline{AB}$ 90° counterclockwise about the origin.

F7 6 (Enter the number of degrees for the rotation.) **ENTER** 90

F5 2 (Move cursor to $\overline{AB}$.) **ENTER** (Move cursor to the origin.) **ENTER** (Move cursor to 90.) **ENTER**

3. Use the software's reflection feature to reflect the result of Step 2 in the y-axis.

F5 4 (Move cursor to the rotated image of $\overline{AB}$.) **ENTER** (Move cursor to the y-axis.) **ENTER**

4. Perform the transformations in reverse order.

Use the software's reflection feature to reflect $\overline{AB}$ in the y-axis.

F5 4 (Move cursor to $\overline{AB}$.) **ENTER** (Move cursor to y-axis.) **ENTER**

5. Rotate the result of Step 4 90° counterclockwise about the origin.

F5 2 (Move cursor to the reflected image of $\overline{AB}$.) **ENTER** (Move cursor to the origin.) **ENTER** (Move cursor to 90.) **ENTER**

Technology Activity Keystrokes

For use with pages 430–436

SKETCHPAD

1. Turn on the axes and the grid. Choose **Snap To Grid** from the **Graph** menu. Choose **Plot Points . . .** from the **Graph** menu. Enter the points $(1, -0.5)$ and $(1.5, -1.5)$, then click OK. Use the text tool to relabel the points A and B, respectively. Draw $\overline{AB}$ using the segment straightedge tool.

2. Choose the translate selection arrow tool. Select the point at the origin and choose **Mark Center** from the **Transform** menu. Select $\overline{AB}$ and its endpoints, then choose **Rotate . . .** from the **Transform** menu. Enter 90 and click OK.

3. Select the y-axis. Choose **Mark Mirror** from the **Transform** menu. Select the rotation of $\overline{AB}$ in Step 2, then choose **Reflect** from the **Transform** menu.

4. Perform the transformations in reverse order. Select $\overline{AB}$, then choose **Reflect** from the **transform** menu.

5. Select the reflection of $\overline{AB}$ in Step 4, then choose **Rotate . . .** from the **Transform** menu. Enter 90 and click OK.

Geometry
Chapter 7 Resource Book

NAME _____ DATE _____

Technology Activity Keystrokes

For use with page 434

Keystrokes for Exercise 28

TI-92

1. Draw a polygon.

   [F3] 4 (Locate first point.) [ENTER] (Move to second point.)

   [ENTER] (Move to third point.) [ENTER] (Repeat for as many points

   as are desired and then return to the first point to close the polygon.)

   [ENTER]

2. Draw a line for reflection.

   [F2] 4 (Locate position for line.) [ENTER] (Move cursor to place

   line.) [ENTER]

3. Draw a vector for translation that is not parallel to the line.

   [F2] 7 (Locate position for tail of vector.) [ENTER] (Move the cursor to

   position for head of vector.) [ENTER]

4. Reflect the polygon with respect to the line.

   [F5] 4 (Place cursor on polygon.) [ENTER] (Move cursor to the line.) [ENTER]

5. Translate the reflected polygon with respect to the vector.

   [F5] 1 (Place cursor on the reflected polygon.) [ENTER] (Move the

   cursor line to the vector.) [ENTER]

6. Translate the original polygon with respect to the vector.

   [F5] 1 (Place cursor on the polygon.) [ENTER] (Move the cursor to

   the vector.) [ENTER]

7. Reflect the translated polygon with respect to the line.

   [F5] 4 (Place cursor on the translated polygon.) [ENTER] (Move

   cursor to line.) [ENTER]

SKETCHPAD

1. Draw a polygon using the segment straightedge tool.

2. Draw a line for reflection using the line straightedge tool.

3. Reflect the polygon with respect to the line (select the line). Choose
   Mark Mirror from the **Transform** menu. Use the selection arrow
   tool to select the segments and points of the polygon. Choose
   Reflect from the **Transform** menu.

4. Translate the reflected polygon. Choose **Translate** from the
   Transform menu. Enter 4 in the magnitude box, Click OK.

5. Reverse the order. Translate the polygon first and then reflect the
   translated polygon.

NAME _____ DATE _____

Practice A

For use with pages 430–436

Match the composition with the diagram.

A.
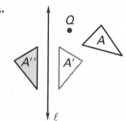

B.
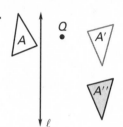

C.
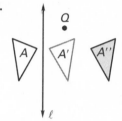

D.

1. Translate parallel to ℓ then reflect in ℓ.

2. Rotate about Q, then translate parallel to ℓ.

3. Rotate about Q, then reflect in ℓ.

4. Reflect in ℓ, then translate perpendicular to ℓ.

Perform the stated transformation on the preimage, $\overline{AB}$. Give the coordinates of the image, $\overline{A'B'}$.

5. Reflection: in the y-axis

6. Rotation: 90° counterclockwise about the origin

7. Translation: $(x, y) \rightarrow (x - 4, y + 3)$

8. Reflection: in $x = -2$

9. Rotation: 180° clockwise about the origin

10. Translation: $(x, y) \rightarrow (x + 5, y - 6)$

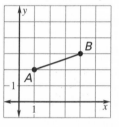

Sketch the image of $A(1, -3)$ after the described glide reflection.

11. **Translation:** $(x, y) \rightarrow (x + 2, y)$
    Reflection: in the x-axis

12. **Translation:** $(x, y) \rightarrow (x - 4, y + 3)$
    Reflection: in $y = 2$

13. **Translation:** $(x, y) \rightarrow (x - 3, y + 2)$
    Reflection: in $x = 2$

14. **Translation:** $(x, y) \rightarrow (x + 5, y - 4)$
    Reflection: in $y = -5$

Describe the composition of the transformations.

15.
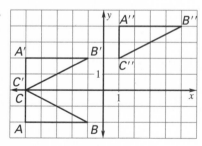

16.
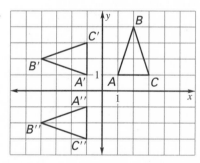

NAME _____ DATE _____

Practice B

For use with pages 430–436

Match the composition with the diagram, in which figure 1 is the preimage of figure 2, and figure 2 is the preimage of figure 3.

A.
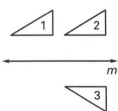

B.
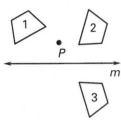

C.
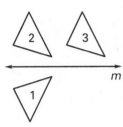

1. Reflect in line *m*, then translate parallel to line *m*.

2. Translate parallel to line *m*, then reflect in line *m*.

3. Rotate about point *P*, then reflect in line *m*.

Perform the stated transformation on the preimage, △*ABC*. Give the coordinates of the image, △*A′B′C′*.

4. Rotation: 90° clockwise about the origin

5. Reflection: in $x = 3$

6. Translation: $(x, y) \rightarrow (x + 3, y - 2)$

7. Rotation: 180° counterclockwise about the origin

8. Translation: $(x, y) \rightarrow (x - 3, y - 2)$

9. Reflection: in the line $y = x$

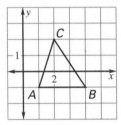

Sketch the image of $A(-4, 2)$ after the described glide reflection.

10. **Translation:** $(x, y) \rightarrow (x, y + 3)$
    Reflection: in the y-axis

11. **Translation:** $(x, y) \rightarrow (x - 2, y + 5)$
    Reflection: in $y = 4$

12. **Translation:** $(x, y) \rightarrow (x - 2, y - 5)$
    Reflection: in $x = -1$

13. **Translation:** $(x, y) \rightarrow (x + 3, y + 2)$
    Reflection: in $y = x$

Describe the composition of the transformations.

14.
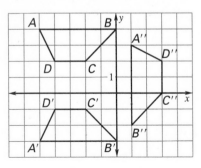

15.
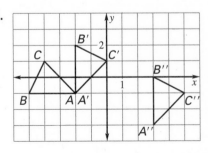

NAME _____ DATE _____

Practice C
For use with pages 430–436

Perform the stated transformation on the preimage, △ABC. Give the coordinates of the image, △A′B′C′.

1. Rotation: 270° clockwise about the origin

2. Reflection: in $x = -2$

3. Translation: $(x, y) \rightarrow (x + 6, y + 4)$

4. Rotation: 90° counterclockwise about the origin

5. Translation: $(x, y) \rightarrow (x - 5, y - 4)$

6. Reflection: in the line $y = -x$

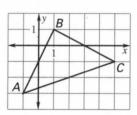

Sketch the image of $A(-3, 5)$ after the described glide reflection.

7. **Translation:** $(x, y) \rightarrow (x + 3, y - 2)$
   Reflection: in the y-axis

8. **Translation:** $(x, y) \rightarrow (x - 6, y)$
   Reflection: in $x = 4$

9. **Translation:** $(x, y) \rightarrow (x + 5, y - 4)$
   Reflection: in $y = -2$

10. **Translation:** $(x, y) \rightarrow (x - 1, y - 4)$
    Reflection: in $y = -x$

Describe the composition of the transformations.

11.

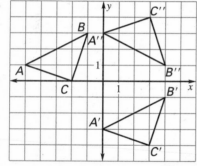

12.

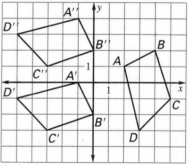

Decide whether the part of the waveform that is below the x-axis is a glide reflection of the part that is above. If it is, write the translation using a vector in component form.

13.

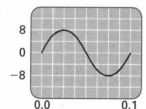

14.

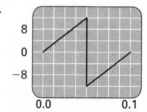

15.

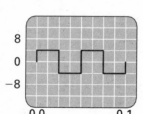

16.

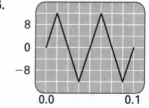

Geometry
Chapter 7 Resource Book

NAME _____ DATE _____

Reteaching with Practice

For use with pages 430–436

GOAL **Identify glide reflections in a plane and represent transformations as compositions of simpler transformations.**

VOCABULARY

A **glide reflection** is a transformation in which every point P is mapped onto a point P'' by the following steps:

1. A translation maps P onto P'.

2. A reflection in a line k parallel to the direction of the translation maps P' onto P''.

When two or more transformations are combined to produce a single transformation, the result is called a **composition** of the transformations.

Theorem 7.6 Composition Theorem
The composition of two (or more) isometries is an isometry.

EXAMPLE 1 *Finding the Image of a Glide Reflection*

Use the information below to sketch the image of $\triangle ABC$ after a glide reflection.

$A(-5, 5)$, $B(-3, 2)$, $C(1, 5)$

Translation: $(x, y) \rightarrow (x, y - 7)$

Reflection: in the line $x = 3$

SOLUTION

Begin by graphing $\triangle ABC$. Then, shift the triangle 7 units down to produce $\triangle A'B'C'$. Finally, reflect the triangle in the line $x = 3$ to produce $\triangle A''B''C''$.

	Coordinates	
$\triangle ABC$	$\triangle A'B'C'$	$\triangle A''B''C''$
$A(-5, 5)$	$A'(-5, -2)$	$A''(11, -2)$
$B(-3, 2)$	$B'(-3, -5)$	$B''(9, -5)$
$C(1, 5)$	$C'(1, -2)$	$C''(5, -2)$

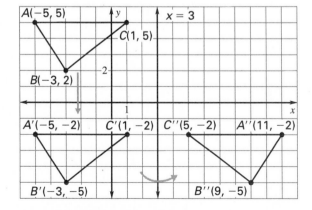

NAME _____ DATE _____

Reteaching with Practice

For use with pages 430–436

Exercises for Example 1

Sketch the image of △*ABC* after a glide reflection using the
given transformations in the order they appear. Then, reverse
the order of the transformations and sketch the image again.
Determine if the order of the transformations affects the image.

1. $A(0, 0), B(0, 5), C(7, 0)$

Translation: $(x, y) \to (x + 3, y)$

Reflection: in the x-axis

2. $A(-3, 2), B(-1, -2), C(3, 2)$

Translation: $(x, y) \to (x - 4, y + 2)$

Reflection: in the line $x = 2$

3. $A(3, -1), B(7, -1), C(6, 2)$

Translation: $(x, y) \to (x - 1, y + 5)$

Reflection: in the line $y = -1$

4. $A(-4, 0), B(0, 7), C(3, 1)$

Translation: $(x, y) \to (x, y + 3)$

Reflection: in the line $x = 4$

EXAMPLE 2 *Finding the Image of a Composition*

Sketch the image of $\overline{AB}$ after a composition of the given rotation and
reflection.

$A(-5, 3), B(-3, 7)$

Rotation: 90° clockwise about the origin

Reflection: in the line $y = 1$

SOLUTION

Begin by graphing $\overline{AB}$. Then rotate the segment
90° clockwise about the origin to produce $\overline{A'B'}$.
Finally, reflect the segment in the line $y = 1$ to
produce $\overline{A''B''}$.

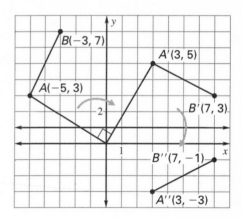

Exercises for Example 2

Sketch the image of $\overline{AB}$ after a composition using the given
transformations.

5. $A(-5, 5), B(-3, 2)$

Translation: $(x, y) \to (x + 8, y - 2)$
Reflection: in the x-axis

6. $A(0, -8), B(3, -4)$

Rotation: 180° clockwise
about the origin

Reflection: in the line $x = 3$

7. $A(6, -1), B(9, 4)$

Translation: $(x, y) \to (x - 8, y + 1)$

Reflection: in the y-axis

8. $A(3, 10), B(7, 5)$

Translation: $(x, y) \to (x - 4, y)$

Rotation: 90° counterclockwise
about the origin

NAME _____ DATE _____

Quick Catch-Up for Absent Students

For use with pages 429–436

The items checked below were covered in class on (date missed) _____

Activity 7.5: Multiple Transformations (p. 429)

____ **Goal:** Determine whether the order in which the two transformations are performed affects the final image.

Lesson 7.5: Glide Reflections and Compositions

____ **Goal 1:** Identify glide reflections in a plane. (p. 430)

Material Covered:

____ Example 1: Finding the Image of a Glide Reflection

Vocabulary:

glide reflection, p. 430

____ **Goal 2:** Represent transformations as compositions of simpler transformations. (pp. 431–432)

Material Covered:

____ Example 2: Finding the Image of a Composition

____ Student Help: Study Tip

____ Example 3: Comparing Orders of Compositions

____ Example 4: Describing a Composition

____ Student Help: Study Tip

____ Example 5: Describing a Composition

Vocabulary:

composition, p. 431

____ Other (specify) _____

_____

Homework and Additional Learning Support

____ Textbook (specify) _pp. 433–436_____

_____

____ *Reteaching with Practice* worksheet (specify exercises) _____

____ *Personal Student Tutor* for Lesson 7.5

NAME _____ DATE _____

Real-Life Application:
When Will I Ever Use This?

For use with pages 430–436

Golf

Many obstacles can exist on a golf course to make it more difficult for a golfer
to obtain a low score. The term used to describe these obstacles is a hazard.
One type of hazard is a water hazard, which can be a pond, stream, or a creek.
Another type of hazard is a sand bunker, which is a depression in the ground
that contains sand. Once a golf ball lands in a sand bunker, the shot to get the
ball out can be extremely difficult.

Suppose you are designing a putting green for one of the holes on a golf course.
In the diagram shown below, you draw bunker *A* first and decide to add three
more bunkers, labeled *B*, *C*, and *D*, that have exactly the same shape as bunker *A*.

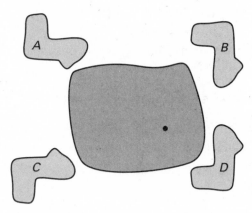

**In Exercises 1–3, describe the transformations that are combined to
obtain the given relation.**

1. bunker *A* to bunker *B*

2. bunker *A* to bunker *C*

3. bunker *A* to bunker *D*

4. Suppose bunker *C* is drawn first. How would you get bunker *D*?

Geometry
Chapter 7 Resource Book

Challenge: Skills and Applications

For use with pages 430–436

In this worksheet, the upper-case letters represent transformations. This notation makes it easy to describe compositions. For example, if T is a translation and U is a rotation, then TU represents the translation T followed by the rotation U, and T^2 represents the translation followed by itself.

In Exercises 1–20, describe the transformation using the notation $(x, y) \rightarrow (?, ?)$, where R, S, T, U, V, and W are transformations defined as follows.

R: translation: $(x, y) \rightarrow (x + 6, y)$ S: translation: $(x, y) \rightarrow (x, y - 4)$

T: translation: $(x, y) \rightarrow (x + 3, y - 2)$ U: 90° clockwise rotation about origin

V: reflection in the x-axis W: reflection in the y-axis

Example: VS

First, V is applied, mapping (x, y) to $(x, -y)$.

Then, S is applied, mapping $(x, -y)$ to $(x, -y - 4)$.

So, VS can be described as the transformation $(x, y) \rightarrow (x, -y - 4)$.

1. T^2	**2.** T^3	**3.** T^4	**4.** T^n
5. RS	**6.** SR	**7.** RV	**8.** VR
9. SW	**10.** VW	**11.** TW	**12.** WT
13. U	**14.** U^2	**15.** U^3	**16.** U^4
17. UV	**18.** VU	**19.** UW	**20.** TU

21. Which of the transformations given in Exercise 1–20 are glide reflections?

22. Two transformations are *equal* if they describe the same mapping. For example, $VW = WV$ because VW and WV both map an arbitrary point (x, y) to $(-x, -y)$. Which of the transformations given in Exercises 1–20 are equal?

In Exercises 23–27, the given transformation can be described as a reflection about line j, followed by a reflection about line k. Find equations for j and k. (Hint: Use Theorem 7.3 and Theorem 7.5. There is more than one correct answer.)

23. translation: $(x, y) \rightarrow (x + 6, y)$ 24. translation: $(x, y) \rightarrow (x, y - 8)$

25. translation: $(x, y) \rightarrow (x - 4, y + 2)$ 26. 180° rotation about origin

27. 90° counterclockwise rotation about origin

Geometry
Chapter 7 Resource Book

81

TEACHER'S NAME _____ CLASS _____ ROOM _____ DATE _____

Lesson Plan

2-day lesson (See *Pacing the Chapter*, TE pages 392C–392D)   For use with pages 437–444

GOALS
1. **Use transformations to classify frieze patterns.**
2. **Use frieze patterns to design border patterns in real life.**

State/Local Objectives _____

_____

✓ Check the items you wish to use for this lesson.

STARTING OPTIONS
____ Homework Check: TE page 433: Answer Transparencies
____ Warm-Up or Daily Homework Quiz: TE pages 437 and 436, CRB page 84, or Transparencies

TEACHING OPTIONS
____ Motivating the Lesson: TE page 438
____ Lesson Opener (Visual Approach): CRB page 85 or Transparencies
____ Examples: Day 1: 1–2, SE pages 437–438; Day 2: 3–4, SE page 439
____ Extra Examples: Day 1: TE page 438 or Transp.; Day 2: TE page 439 or Transp.
____ Closure Question: TE page 439
____ Guided Practice: SE page 440   Day 1: Exs. 1–7; Day 2: none

APPLY/HOMEWORK
Homework Assignment
____ Basic   Day 1: 8–26; Day 2: 29–45, 49–52, 54–64; Quiz 2: 1–7
____ Average   Day 1: 8–26; Day 2: 29–52, 54–64; Quiz 2: 1–7
____ Advanced   Day 1: 8–26; Day 2: 29–64; Quiz 2: 1–7

Reteaching the Lesson
____ Practice Masters: CRB pages 86–88 (Level A, Level B, Level C)
____ Reteaching with Practice: CRB pages 89–90 or Practice Workbook with Examples
____ Personal Student Tutor

Extending the Lesson
____ Applications (Interdisciplinary): CRB page 92
____ Challenge: SE page 443; CRB page 93 or Internet

ASSESSMENT OPTIONS
____ Checkpoint Exercises: Day 1: TE page 438 or Transp.; Day 2: TE page 439 or Transp.
____ Daily Homework Quiz (7.6): TE page 444, or Transparencies
____ Standardized Test Practice: SE page 443; TE page 444; STP Workbook; Transparencies
____ Quiz (7.4–7.6): SE page 444; CRB page 94

Notes _____

_____

_____

TEACHER'S NAME _____ CLASS _____ ROOM _____ DATE _____

Lesson Plan for Block Scheduling

1-day lesson (See *Pacing the Chapter,* TE pages 392C–392D**)**   For use with pages 437–444

GOALS
1. **Use transformations to classify frieze patterns.**
2. **Use frieze patterns to design border patterns in real life.**

State/Local Objectives _____

_____

_____

✓ **Check the items you wish to use for this lesson.**

STARTING OPTIONS

____ Homework Check: TE page 433: Answer Transparencies

____ Warm-Up or Daily Homework Quiz: TE pages 437 and
      436, CRB page 84, or Transparencies

CHAPTER PACING GUIDE	
Day	**Lesson**
1	7.1 (all)
2	7.2 (all)
3	7.3 (all)
4	7.4 (all); 7.5 (begin)
5	7.5 (end); **7.6 (begin)**
6	**7.6 (end)**; Review Ch. 7
7	Assess Ch. 7; 8.1 (all)

TEACHING OPTIONS

____ Motivating the Lesson: TE page 438

____ Lesson Opener (Visual Approach): CRB page 85 or Transparencies

____ Examples: Day 5: 1–2, SE pages 437–438; Day 6: 3–4, SE page 439

____ Extra Examples: Day 5: TE page 438 or Transp.; Day 6: TE page 439 or Transp.

____ Closure Question: TE page 439

____ Guided Practice: SE page 440   Day 5: Exs. 1–7; Day 6: none

APPLY/HOMEWORK

Homework Assignment (See also the assignment for Lesson 7.5.)

____ Block Schedule: Day 5: 8–26; Day 6: 29–48, 50–55, 57–67; Quiz 2: 1–7

Reteaching the Lesson

____ Practice Masters: CRB pages 86–88 (Level A, Level B, Level C)

____ Reteaching with Practice: CRB pages 89–90 or Practice Workbook with Examples

____ Personal Student Tutor

Extending the Lesson

____ Applications (Interdisciplinary): CRB page 92

____ Challenge: SE page 443; CRB page 93 or Internet

ASSESSMENT OPTIONS

____ Checkpoint Exercises: Day 5: TE page 438 or Transp.; Day 6: TE page 439 or Transp.

____ Daily Homework Quiz (7.6): TE page 444, or Transparencies

____ Standardized Test Practice: SE page 443; TE page 444; STP Workbook; Transparencies

____ Quiz (7.4–7.6): SE page 444; CRB page 94

Notes _____

_____

_____

NAME _____ DATE _____

WARM-UP EXERCISES

For use before Lesson 7.6, pages 437–444

Match the transformation of the preimage with its image.

1. translation

2. rotation

3. reflection

4. glide reflection

preimage   A   B

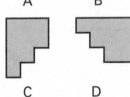

C   D

. . . .

DAILY HOMEWORK QUIZ

For use after Lesson 7.5, pages 429–436

1. Name the coordinates of the image of $A(-3, 2)$ after a glide reflection.

    Translation: $(x, y) \rightarrow (x + 4, y)$

    Reflection: in the x-axis

2. Name the coordinates of the image of $A(-3, 2)$ after a composition of the given rotation and reflection.

    Rotation: 90° counterclockwise about the origin

    Reflection: in the y-axis

3. Describe the composition of the transformations.

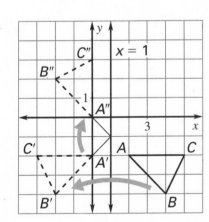

NAME _____ DATE _____

Visual Approach Lesson Opener

For use with pages 437–444

YOU WILL NEED: • graph paper • colored pencils or markers

1. The pattern below was designed for a three-color hand-knit headband. The pattern extends to the left and to the right in such a way that the pattern can be mapped onto itself by a horizontal translation. Each square in the graph paper represents one stitch of the color shown. The width of the headband is 19 rows (as shown), and the length of the headband is 84 stitches (not all shown, because the pattern repeats horizontally). How many stitches are knit in each row until the pattern repeats? How many times does the repeating pattern appear in the full length of the headband? Identify any translations, rotations, reflections, or glide reflections you see in the headband.

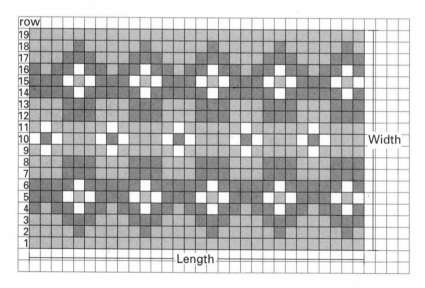

2. Design your own pattern for a hand-knit headband. Make sure your pattern extends to the left and to the right in such a way that the pattern can be mapped onto itself by a horizontal translation. Use 2–5 colors and 15–20 rows. How many stitches are knit in each row until your pattern repeats? Identify any translations, rotations, reflections, or glide reflections in your headband.

Practice A

For use with pages 437–444

Fill in the blanks with the abbreviation which matches the type of isometry.

1. _____ Translation, horizontal line reflection, and glide reflection

2. _____ Translation, 180° rotation, vertical line reflection, and glide reflection

3. _____ Translation and vertical line reflection

4. _____ Translation only

5. _____ Translation, 180° rotation, horizontal line reflection, vertical line reflection, and glide reflection

6. _____ Translation and 180° rotation

7. _____ Translation and glide reflection

Name the isometries that map the frieze pattern onto itself.

8.

9.

10.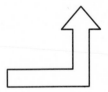

11.

Use the design below to create a frieze pattern with the given classification.

12. TV

13. T

14. TG

15. TR

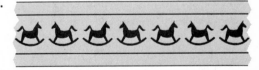

Use the chart on page 438 to classify the frieze pattern.

16.

17.

18.

19.

NAME _____ DATE _____

Practice B

For use with pages 437–444

Use the diagram of the frieze pattern.

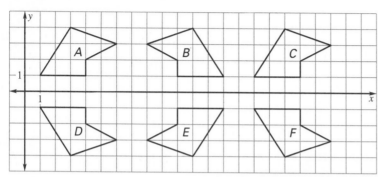

1. Is there a reflection in a horizontal line? If so, describe the reflection(s).

2. Is there a reflection in a vertical line? If so, describe the reflection(s).

3. Is there a 180° rotation? If so, describe the point(s) of rotation.

4. Name and describe the transformation that maps A onto C.

5. Name and describe the transformation that maps A onto D.

6. Name and describe the transformation that maps A onto E.

7. Name and describe the transformation that maps A onto F.

Use the design below to create a frieze pattern with the given classification.

8. TV   9. TG

10. TRVG   11. THG

Use the chart on page 438 to classify the frieze pattern.

12.

13.

14.

15.

NAME _____ DATE _____

Practice C

For use with pages 437–444

Use the diagram of the frieze pattern.

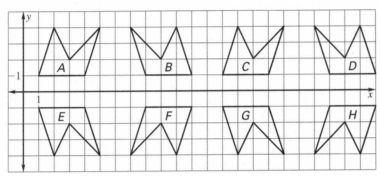

1. Is there a reflection in a horizontal line? If so, describe the reflection(s).

2. Is there a reflection in a vertical line? If so, describe the reflection(s).

3. Is there a 180° rotation? If so, describe the point(s) of rotation.

4. Name and describe the transformation that maps *A* onto *C*.

5. Name and describe the transformation that maps *A* onto *E*.

6. Name and describe the transformation that maps *A* onto *F*.

7. Name and describe the transformation that maps *A* onto *G*.

Use the design below to create a frieze pattern with the given classification.

8. TR

9. TG

10. THG

11. TRVG

Use the chart on page 438 to classify the frieze pattern.

12.

13.

14.

15.

Lesson 7.6

NAME _____ DATE _____

Reteaching with Practice

For use with pages 437–443

GOAL Use transformations to classify frieze patterns.

VOCABULARY

A **frieze pattern** or **border pattern** is a pattern that extends to the left and right in such a way that the pattern can be mapped onto itself by a horizontal translation.

Classification of Frieze Patterns

T	Translation
TR	Translation and 180° rotation
TG	Translation and horizontal glide reflection
TV	Translation and vertical line reflection
THG	Translation, horizontal line reflection, and horizontal glide reflection
TRVG	Translation, 180° rotation, vertical line reflection, and horizontal glide reflection
TRHVG	Translation, 180° rotation, horizontal line reflection, vertical line reflection, and horizontal glide reflection

---

EXAMPLE 1 *Classifying Patterns*

Name the isometries that map the frieze pattern onto itself.

a.   b.

SOLUTION

a. This frieze pattern can be mapped onto itself by a horizontal translation (T).

b. This frieze pattern can be mapped onto itself by a horizontal translation (T)

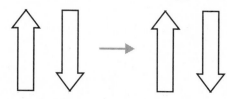

or by a horizontal glide reflection (G).

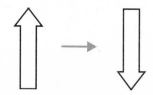

NAME _____ DATE _____

Reteaching with Practice

For use with pages 437–444

Exercises for Example 1

In Exercises 1–5, name the isometries that map the frieze pattern onto itself.

1.  2.

3.

EXAMPLE 2 **Describing Transformations**

Use the diagram of the frieze pattern.

a. Is there a reflection in a vertical line?

b. Is there a reflection in a horizontal line?

c. Name and describe the transformation that maps A onto F.

d. Name and describe the transformation that maps D onto E.

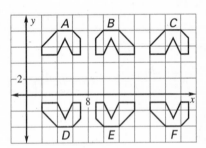

SOLUTION

a. Yes, there is a reflection in the line $x = 8$ and also in the line $x = 15$.

b. Yes, there is a reflection in the line $y = 2$.

c. A can be mapped onto F by a horizontal glide reflection.

d. D can be mapped onto E by a translation.

Exercises for Example 2

In Exercises 4–7, use the diagram of the frieze pattern.

4. Is there a reflection in a horizontal line? If so, describe the reflection(s).

5. Is there a reflection in a vertical line? If so, describe the reflection(s).

6. Name and describe the transformation that maps B onto C.

7. Name and describe the transformation that maps D onto C.

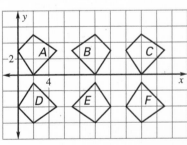

Quick Catch-Up for Absent Students

For use with pages 437–444

The items checked below were covered in class on (date missed) _____

Lesson 7.6: Frieze Patterns

____ **Goal 1:** Use transformations to classify frieze patterns. (pp. 437–438)

Material Covered:

____ Example 1: Describing Frieze Patterns

____ Student Help: Study Tip

____ Example 2: Classifying a Frieze Pattern

Vocabulary:

frieze pattern, p. 437   border pattern, p. 437

____ **Goal 2:** Use frieze patterns to design border patterns in real life.

Material Covered:

____ Example 3: Identifying Frieze Patterns

____ Example 4: Drawing a Frieze Pattern

____ Other (specify) _____

_____

Homework and Additional Learning Support

____ Textbook (specify) _pp. 440–444_ _____

_____

____ *Reteaching with Practice* worksheet (specify exercises)_____

____ *Personal Student Tutor* for Lesson 7.6

NAME _____ DATE _____

Interdisciplinary Application

For use with pages 437–444

Musical Notation

MUSIC   A musical staff is the component of musical notation that consists of five horizontal lines on which notes are written as shown below. The symbol at the beginning of the staff is called the G clef, or treble clef. The clef sign determines the pitch of the notes.

musical staff

A note is a symbol that when placed on the musical staff indicates a tone that is to be played for a relative length of time. A rest is a character that indicates a rhythmic silence of a relative duration of time. Some of the common symbols of musical terminology are shown below.

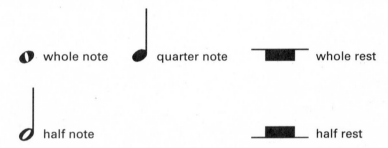

whole note   quarter note   whole rest

half note   half rest

In Exercises 1 and 2, determine whether the combination of notes and/or rests form a frieze pattern. If so, classify the frieze pattern.

1.

2.

3. The stem on a note appears on the right side of the head when turned up, but on the left side when turned down (J Γ). Following this rule, when notes have a frieze pattern, what classifications would they not be able to meet? Explain.

Challenge: Skills and Applications

For use with pages 437–444

As you know, a frieze pattern is a pattern that can be mapped onto itself by a horizontal translation. A *wallpaper pattern* covers the plane and can be mapped onto itself by translations in more than one direction.

Just as there are 7 categories of frieze patterns, there are 17 categories of wallpaper patterns, depending on the kinds of symmetry in the pattern.

In Exercises 1–6, assume that the wallpaper pattern extends indefinitely. Name the isometries that map the wallpaper pattern onto itself.

Example:

The isometries for this pattern are translation, 120° rotation, reflection in a vertical line, vertical glide reflection, reflection in a line that makes a 30° angle with the horizontal, and glide reflection along a line that makes a 30° angle with the horizontal.

1.

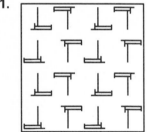

2.

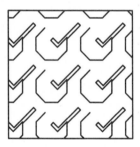

3.

4.

5.

6.

NAME _____ DATE _____

Write the coordinates of the vertices A', B', **and** C' **after**
$\triangle ABC$ **is translated by the given**
vector. *(Lesson 7.4)*

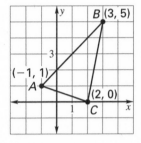

1. $\langle 1, 2 \rangle$ **2.** $\langle -2, 3 \rangle$

3. $\langle -1, -2 \rangle$ **4.** $\langle 4, 3 \rangle$

In Exercises 5 and 6, sketch the image of $\triangle ABC$ **after a**
composition using the given transformations in the order
they appear. *(Lesson 7.5)*

5. $A(-1, 1)$, $B(-4, 4)$, $C(-5, 1)$ **6.** $A(1, 1)$, $B(-3, 2)$, $C(-1, 4)$

    Translation: **Translation:**

    $(xy) \rightarrow (x - 1, y - 5)$ $(x, y) \rightarrow (x - 2, y - 3)$

    Reflection: in the y-axis   **Rotation:** 90° counterclockwise
                                         around origin

7. *Musical Notes*   Do the notes shown form a frieze pattern? If so,
    classify the frieze pattern. *(Lesson 7.6)*

Answers

1. _____

_____

_____

2. _____

_____

_____

3. _____

_____

_____

4. _____

_____

_____

5. _____

6. _____

7. _____

_____

_____

_____

_____

Chapter Review Games and Activities

For use after Chapter 7

Complete the following crossword puzzle.

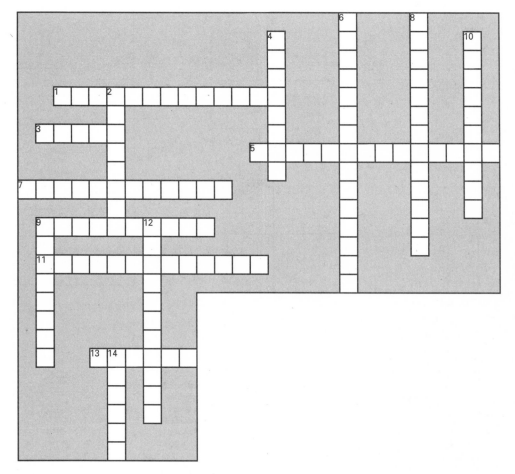

Across

1. One way of representing a vector.

3. The new figure after a transformation.

5. Describes how geometric figures of the same shape are related to one another.

7. Starting point of a vector.

9. A transformation that uses a line that acts like a mirror.

11. Ending point of a vector.

13. Quantity that has both direction and magnitude.

Down

2. The original figure before a transformation.

4. A transformation that preserves length.

6. A translation followed by a reflection.

8. Can be mapped onto itself by a horizontal translation.

9. A transformation in which a figure is turned about a fixed point.

10. The center of rotation.

12. A transformation whereby a figure glides onto its image.

14. A Dutch graphic artist whose works include optical illusions and geometric patterns.

Review and Assess

NAME _____ DATE _____

Chapter Test A

For use after Chapter 7

Classify the transformation as a *reflection, rotation,* or *translation*.

1.
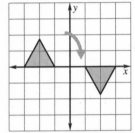

2.
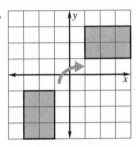

3.
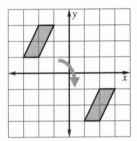

4.
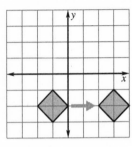

1. _____

2. _____

3. _____

4. _____

5. _____

6. _____

7. _____

8. _____

9. _____

10. _____

11. See left. _____

How many lines of symmetry does each flag have?

5. Venezuela

6. Canada

7. United States

Use the diagram.

8. Identify the transformation that maps figure *ABCDE* onto figure *STUVX*.

9. What is the preimage of *V*?

10. What is the image of *A*?

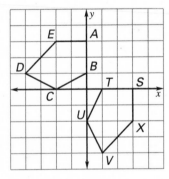

11. Sketch a polygon that has line symmetry but not rotational symmetry.

12. Sketch a polygon that has both line symmetry and rotational symmetry.

12.	See left.
13.	
14.	
15.	
16.	
17.	
18.	
19.	See left.
20.	See left.

Name the vector and write its component form.

13.

14.

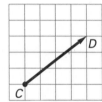

Use the figure to match the translation of □ABCD to □A′B′C′D′ by using the given vector.

A. $A'(-2, 1), B'(2, 1),$
$C'(1, -2), D'(-3, -2)$

B. $A'(-5, -1), B'(-1, -1),$
$C'(-2, -4), D'(-6, -4)$

C. $A'(-4, 3), B'(0, 3),$
$C'(-1, 0), D'(-5, 0)$

D. $A'(-3, 0), B'(1, 0),$
$C'(0, -3), D'(-4, -3)$

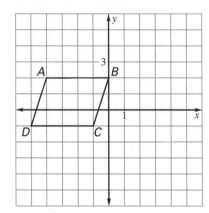

15. $\vec{u} = \langle 0, 1 \rangle$ vector

16. $\vec{u} = \langle 2, -1 \rangle$ vector

17. $\vec{u} = \langle 1, -2 \rangle$ vector

18. $\vec{u} = \langle -1, -3 \rangle$ vector

Name all of the isometrics that map the frieze patterns onto itself.

19.

20.

NAME _____ DATE _____

Chapter Test B

For use after Chapter 7

Classify the transformation as a *reflection, rotation,* or *translation*.

1.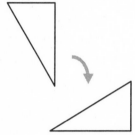

2.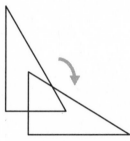

3.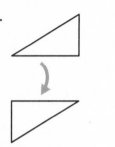

4.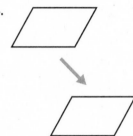

1. _____

2. _____

3. _____

4. _____

5. _____

6. _____

7. _____

8. _____

9. _____

10. _____

11. See left. _____

How many lines of symmetry does each flag have?

5. Ireland

6. Iraq

7. Puerto Rico

Use the diagram.

8. Identify the transformation that maps figure *DEFGHIJK* onto figure *MNOPQRST*.

9. What is the preimage of *S*?

10. What is the image of *D*?

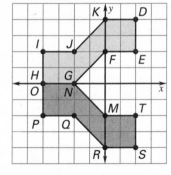

11. Sketch a polygon that has line symmetry but not rotational symmetry.

12. Sketch a polygon that has both line symmetry and rotational symmetry.

Name the vector and write its component form.

13.

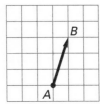

14.

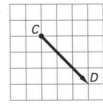

Use the figure to match the translation of ▱*ABCD* to ▱*A′B′C′D′* by using the given vector.

A. $A'(0, 5), B'(5, 7),$
  $C'(8, 6), D'(3, 3)$

B. $A'(3, 3), B'(8, 5),$
  $C'(11, 4), D'(6, 1)$

C. $A'(1, 1), B'(6, 3),$
  $C'(9, 2), D'(4, -1)$

D. $A'(-1, 1), B'(4, 3),$
  $C'(7, 2), D'(2, -1)$

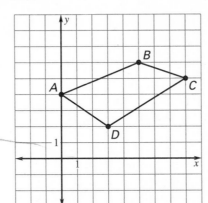

15. $\vec{u} = \langle 1, -3 \rangle$

16. $\vec{u} = \langle 0, 1 \rangle$

17. $\vec{u} = \langle -1, -3 \rangle$

18. $\vec{u} = \langle 3, -1 \rangle$

Name all of the isometries that map the frieze patterns onto itself.

19.

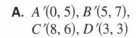

20.

Review and Assess

NAME _____ DATE _____

Chapter Test C

For use after Chapter 7

Classify the transformation as a *reflection, rotation,* or *translation*.

1.

2.

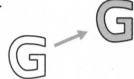 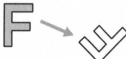

3.

4.

1.	_____
2.	_____
3.	_____
4.	_____
5.	_____
6.	_____
7.	_____
8.	_____
9.	_____
10.	_____
11.	See left. _____

How many lines of symmetry does each flag have?

5. Hungary

6. Tanzania

7. Japan

Use the diagram.

8. Identify the transformation that maps figure *DEFGHIJ* onto figure *MNPQRST*.

9. What is the preimage of *S*?

10. What is the image of *D*?

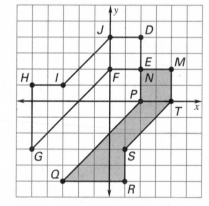

11. Sketch a polygon that has line symmetry but not rotational symmetry.

12. Sketch a polygon that has rotational symmetry but not line symmetry.

Name the vector and write its component form.

13.

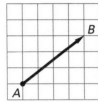

14.

12. See left. _____

13. _____

14. _____

15. _____

16. _____

17. _____

18. _____

19. See left. _____

20. See left. _____

Use the figure to match the translation of □ABCD to □A'B'C'D' by using the given vector.

A. $A'(-4, 3), B'(3, 6),$
  $C'(6, 3), D'(-1, 0)$

B. $A'(-1, 1), B'(6, 4),$
  $C'(9, 1), D'(2, -2)$

C. $A'(-5, -1), B'(2, 2),$
  $C'(5, -1), D'(-2, -4)$

D. $A'(-3, -1), B'(4, 2),$
  $C'(7, -1), D'(0, -4)$

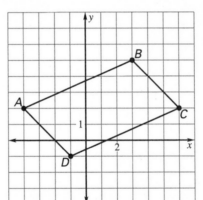

15. $\vec{u} = \langle 1, -3 \rangle$ **16.** $\vec{u} = \langle 0, 1 \rangle$

17. $\vec{u} = \langle -1, -3 \rangle$ **18.** $\vec{u} = \langle 3, -1 \rangle$

Name all of the isometries that map the frieze patterns onto itself.

19.

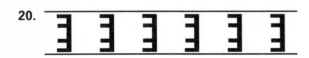

20.

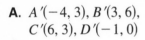

Review and Assess

1. The point $P(-3, 7)$ is reflected in the line $y = x$. What are the coordinates of P'?

   A $(-3, -7)$ **B** $(3, -7)$

   C $(3, 7)$ **D** $(-7, 3)$

   E none of these

2. What two transformations were performed to obtain $\triangle X''Y''Z''$ in the diagram?

   A reflection, reflection

   B reflection, rotation

   C reflection, translation

   D translation, reflection

   E rotation, reflection

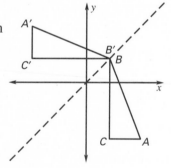

3. What type of transformation is shown in the diagram?

   A slide

   B translation

   C reflection

   D rotation

   E isometry

4. How many lines of symmetry does the polygon at the right have?

   A 2

   B 10

   C 4

   D 8

   E 6

5. Which of the following is *not* a rotation of the figure at the right?

   A   **B**

   C **D**

   E

6. How many lines of symmetry does the infield of a baseball field have?

   A 0 **B** 1

   C 2 **D** 3

   E 4

7. Name all the isometries that will map the snakeskin frieze pattern onto itself.

   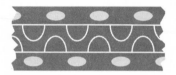

   A translation and rotation

   B vertical line reflection

   C glide reflection

   D only A & B

   E all of the above

Alternative Assessment and Math Journal

For use after Chapter 7

JOURNAL **1.** Using the right triangle shown create a
frize pattern with the given classification.
(a) T, (b) TV, and (c) TR.

MULTI-STEP PROBLEM **2.** Copy the diagram on a coordinate plane.

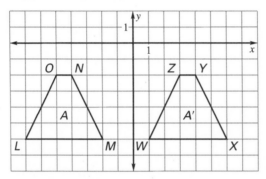

    a. Identify the transformation $LMNO \rightarrow WXYZ$ using coordinate
notation.

    b. What is the preimage of $\overline{XY}$?

    c. Graph the reflection of figure A' in the x-axis. Label the reflection A''.

    d. Identify the coordinates of the vertices of A''.

    e. Rotate A'' 90° counterclockwise about the origin. Label the rota-
tion A'''.

3. *Critical Thinking*   Use your diagram from Exercise 2 to answer the
following.

    a. Find *two* transformations that will return A''' to A. (*Hint:*
Reversing the above transformations will not work because that
requires *three* transformations.)

    b. The transformations from A to A'' are known as a glide reflection.
Using a coordinate plane to show your work, reverse the order of
the glide reflection. Does the resulting image have the same coordinates as A''?

    c. The transformations from A' to A''' are a reflection and a rotation.
Reverse the order of the transformations. Does the order appear to
affect the final image? Show your work in a coordinate plane.

4. *Writing*   Write a paragraph proof for Case 2
of Theorem 7.1. In Case 2, it is given that a
reflection in m maps P onto P' and Q onto Q'.
Prove: $PQ = P'Q'$.

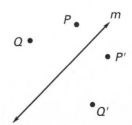

Alternative Assessment Rubric

For use after Chapter 7

JOURNAL
SOLUTION

1. Here are some sample patterns:

a.

b.

c.

MULTI-STEP
PROBLEM
SOLUTION

2. **a.** $(x, y) \rightarrow (x + 8, y)$

b. $\overline{MN}$

c. Check graph. (See Answers beginning on page A1 for graphs.)

d. $(3, 2), (4, 2), (6, 6), (1, 6)$

e. Check graph. (See Answers beginning on page A1 for graphs.)

3. **a.** Rotate A''' 90° counterclockwise about the origin. Translate A'''
$(x, y) \rightarrow (x - 1, y)$.

b. Check graph. (See Answers beginning on page A1 for graphs.). Yes.

c. No, check graph. (See Answers beginning on page A1 for graphs.).

4. Answers may vary. Students' final statement should be $PQ = P'Q'$.

MULTI-STEP
PROBLEM
RUBRIC

4 Students answer all parts of the problem correctly, showing work in a step-by-step manner. Students' graphs are correct and clearly labeled. Students' explanations are correct and clear. Students' proofs are correct and follow a logical order.

3 Students complete all parts of the problem. Students' graphs are complete, but may contain a labeling error. Students' explanations are complete. Students' proofs are complete.

2 Students complete all parts of the problem. Students' graphs are complete, but the order of the transformations may be incorrect. Students' explanations are complete, but they do not find two transformations that will return A''' to A. Students' proofs are complete, but do not follow a logical order.

1 Students' answers are incomplete. Students' graphs are incomplete. Students' explanations do not find two transformations that will return A''' to A. Students' proofs are incomplete and do not prove the final statement.

Project: Designing a Quilt Sample

For use with Chapter 7

OBJECTIVE Design quilt patterns using transformations.

MATERIALS square paper, several pieces of graph paper, ruler, colored pencils or markers, poster board, tape or glue

INVESTIGATION Throughout the centuries, patterns involving color and shape have been present in clothing, household items, art, and architecture. Follow Rules 1–5 below to design one pattern square. Then follow the directions to create at least three quilt samples by applying transformations to that pattern square.

Design Rules for the Pattern Square   Read carefully before starting.

1. Use your ruler to mark the midpoints of each side of a square piece of paper.

2. Draw a line segment, beginning at either a midpoint of a side or a vertex of the square and ending at a midpoint or vertex that is not on the same side of the square.

3. Continue drawing segments from either a midpoint of a segment you have drawn or any vertex. You should follow these rules:

   • You must not cross a segment you previously drew (stop when your new line reaches it).

   • At least one of the new regions you create when drawing each new segment must be a triangle.

4. Draw segments in a similar manner until you have created 8 regions.

5. Color your pattern square.

Creating the Samples   A quilt design is typically created when a pattern square is used repetitively to fill a particular space, such as a quilt to cover a full size bed. Your quilt samples will be smaller than that. You will cover one sheet of graph paper for each sample.

Make a copy of your original pattern square on a small piece of graph paper. Use this as your template to create three different quilt samples. For each sample, use a different full-size sheet of graph paper and apply transformations to the template to fill the page. Color each sample in the same manner as the original pattern square.

PRESENT YOUR RESULTS Prepare a poster that contains your three quilt samples and identifies the transformations you used for each. Write a report detailing how you created the original pattern square. Include the small template. Describe in more detail the transformations you used for each quilt sample and the type of symmetry shown, if any. Tell which quilt sample you like the best and why. Also, tell how your understanding of transformations and artistic design has improved.

Review and Assess

Geometry
Chapter 7 Resource Book

105

Project: Teacher's Notes

For use with Chapter 7

GOALS
- Use transformations in real-life applications
- Identify types of symmetry

MANAGING THE PROJECT The first pattern square can be constructed with plain paper. Paper that is $8\frac{1}{2}$ in. by 11 in. lends itself naturally to squares that are either $8\frac{1}{2}$ inches or $4\frac{1}{4}$ inches on a side. Quilt samples may be done on plain paper, but usually graph paper is easier to use. If you use quarter-inch graph paper, you may want to specify that students create a 2-inch by 2-inch square template. Suggest that the first design be drawn in pencil so changes can be made. You might want to have students work on the initial design in class, having students check each other's work to be sure that the rules are followed. Posters can be displayed in the class-room or somewhere else in the school.

RUBRIC **The following rubric can be used to assess student work.**

4 The student completes three original, creative quilt samples which apply transformations, and presents them neatly. The student includes the small square template and correctly identifies the transformations and the types of symmetry illustrated in the samples. The descriptions of how the original pattern square was created and how transformations were used is detailed and complete. A clear, thoughtful self-assessment of the design experience is included.

3 The student completes three different quilt samples which apply transformations and includes the small square template. However, the template or samples may not be neatly drawn and colored. The student correctly identifies most of the transformations and the types of sym-metry illustrated in the samples. The descriptions of how the original pattern square was created and how transformations were used may be unclear in places. A written self-assessment of the design experience is included, but may not show as much thought as desired.

2 The student completes at least two different quilt samples which apply transformations. The third sample may not actually include a transfor-mation of the original pattern square. The designs may not be neatly drawn and colored or the report may not include the small square template. The student may not correctly identify the transformations and the types of symmetry in the samples. The description of how the original pattern square was created, the descriptions of how transforma-tions were used, and the self-assessment of the design experience are included, but ideas may be sketchy or unclear.

1 The student completes one quilt sample which applies transformations to the small square template, but it is not neat. Transformations or types of symmetry in the sample are not identified or the identification shows a lack of understanding of basic concepts. Descriptions of the design process and the self-assessment are missing or incomplete.

Cumulative Review

For use after Chapters 1–7

Name a point that is collinear with the given points. (1.1)

1. *A* and *C*

2. *G* and *F*

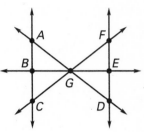

Complete the statement given that $m\angle AGB = m\angle DGB = 90°$. **(2.6)**

3. If $m\angle CGD = 60°$, then the $m\angle BGC =$ __?__ .

4. If $m\angle FGA = 60°$, then the $m\angle AGC =$ __?__ .

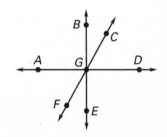

Find the value of *x* that makes *l* ∥ *m*. (3.4)

5.

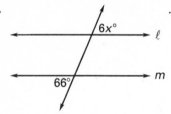

6.

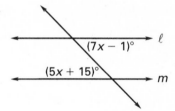

Prove the following using a two-column format. (4.3 and 4.4)

7. Given: $\overline{AB} \parallel \overline{XY}$

       O is a midpoint of $\overline{BY}$.

Prove: $\triangle ABO \cong \triangle XYO$

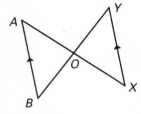

8. Given: $\angle A \cong \angle D$

       $\angle B \cong \angle C$

Prove: $\overline{AB} \cong \overline{DC}$

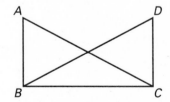

Find the value of *x*. (6.4)

9. *JKLM* is a rectangle.

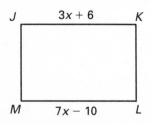

10. *TCBY* is a rhombus.

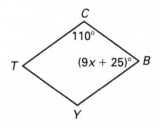

Review and Assess

Use the diagrams to complete the statements. (7.1)

11. $\angle C \cong$ ___?___

12. $\overline{BC} \cong$ ___?___

13. $\overline{DA} \cong$ ___?___

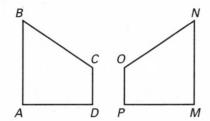

Decide whether the conclusion is *true* or *false*. (7.2)

14. If $A(4, 3)$ is reflected in the line $y = 2$, then A' is $(4, 1)$.

15. If $W(-3, 6)$ is reflected in the line $y = x$, then W' is $(3, 6)$.

The diagonals of the regular hexagon below form six equilateral triangles. Use the diagram to complete the sentence. (7.3)

16. A clockwise rotation of $60°$ about P maps M on to ___?___.

17. A counterclockwise rotation of $180°$ about P maps ___?___ on to N.

18. A clockwise rotation of $60°$ about J maps K on to ___?___.

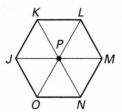

The image of $\triangle ABC$ after a translation is shown below. Use the vector that describes the translation to give the vertices of the preimage. (7.4)

19. $\overrightarrow{PQ} = \langle 3, -1 \rangle$

20. $\overrightarrow{PQ} = \langle -2, 2 \rangle$

21. $\overrightarrow{PQ} = \langle -1, -2 \rangle$

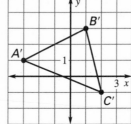

Sketch the image of $\overline{AB}$ after a composition of the given transformations in the order in which they appear. (7.5)

22. $A(2, 3)$, $B(6, 6)$ *Reflection:* in the y axis;
Rotation: $90°$ counterclockwise about the origin

23. $A(-4, 1)$, $B(1, 2)$ *Translation:* $(x, y) \rightarrow (x - 2, y + 1)$;
Reflection: about the x-axis

24. $A(-3, 4)$, $B(-1, -2)$ *Rotation:* $90°$ about the origin; *Translation:* $(x, y) \rightarrow$
$(x + 3, y + 1)$

Use the design below to create a frieze pattern with the given classification. (7.6)

25. TR

26. TRVG

Review and Assess

ANSWERS

Chapter Support

Parent Guide
7.1: reflection   **7.2:** $A'(1, -3), B'(-1, -1),$ $C'(-3, -2)$ **7.3:** 30°   **7.4:** $A'(3, -1),$ $B'(1, -3), C'(-1, -2)$ **7.5:** $A'(2, -5),$ $B'(4, -2), C'(2, -2)$; a reflection in the x-axis and in the y-axis; no   **7.6:** TV

Prerequisite Skills
1. not congruent   **2.** congruent   **3.** congruent
4. not congruent   **5.** congruent   **6.** not
congruent   **7.** $\overline{MN}$ **8.** 40°   **9.** 30°   **10.** $\overline{ON}$
11. 8.2   **12.** $\angle J$ **13.** 110°   **14.** 70°   **15.** 70°
16. 70°   **17.** 180°   **18.** 110°

Strategies for Reading Mathematics
1. $\overline{M'N'}$; M' is the image of M and N' is the image of N.   **2.** $\overline{N'P'}$; N' is the image of N and P' is the image of P.   **3.** $\overline{PM}$; P is the preimage of P' and M is the preimage of M'.

4. $\triangle RST \rightarrow \triangle R'S'T'$

Lesson 7.1

Warm-Up Exercises
1. $\angle X \cong \angle M, \angle Y \cong \angle N, \angle Z \cong \angle O,$
$\overline{XY} \cong \overline{MN}, \overline{YZ} \cong \overline{NO},$ and $\overline{XZ} \cong \overline{MO}$ **2.** 73°
3. $(2, -5), (-2, -5),$ and $(-2, 5)$ **4.** $(1, 3)$

Daily Homework Quiz
1. 312 sq. units   **2.** 56 sq. units
3. 192 sq. units   **4.** 184 sq. units

Lesson Opener
Allow 15 minutes.

1.
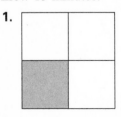

2. *Sample answer:* Translate the square three times: (1) 0 in. horizontally and 1 in. vertically; (2) 0 in. horizontally and 1 in. vertically; (3) 0 in. horizontally and 1 in. vertically.

3.   rhombus
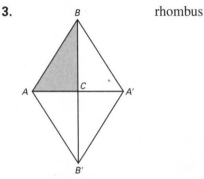

4.   parallelogram
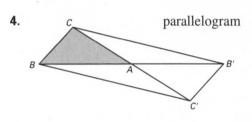

Practice A
1. translation   **2.** rotation   **3.** reflection
4. non-rigid transformation   **5.** rotation
6. non-rigid transformation   **7.** *FGHIJ*
8. reflection over the y-axis   **9.** $\overline{HI}$ **10.** $\overline{AE}$
11. $(2, 3)$ **12.** $DE = 2\sqrt{2}; IJ = 2\sqrt{2}$
13. *JKL*   **14.** *PQR*   **15.** *JLK*   **16.** *LKJ*
17. *FED*   **18.** *QRP*   **19.**   **20.**
21. Answers vary.

Practice B
1. reflection   **2.** translation   **3.** rotation
4. *GHIJKL*   **5.** rotation about the origin   **6.** $\overline{IJ}$
7. $\overline{BC}$ **8.** $(3, 4)$ **9.** $EF = \sqrt{5}; KL = \sqrt{5}$
10. translation or reflection   **11.** reflection
12. rotation   **13.** translation or reflection
14. *FED*   **15.** *BCA*   **16.** $a = 73, b = 53,$ $c = 15, d = 8$ **17.** $p = 19, q = 3, r = 7.5$

Lesson 7.1 *continued*

Practice C

1. rotation   2. reflection   3. translation

4. *EFGH*   5. rotation about the origin

6. $\overline{GH}$ 7. $\overline{BC}$ 8. $(5, 1)$

9.

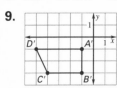

10. $AB = \sqrt{17}$,
$BC = \sqrt{13}$, $CA = \sqrt{10}$,
$DE = \sqrt{17}$, $EF = \sqrt{13}$,
$FD = \sqrt{10}$

11. $PQ = 5$, $QR = 6$, $RP = \sqrt{13}$, $MN = 5$,
$NO = 6$, $OM = \sqrt{13}$ 12. $x = 7$, $y = 4$, $z = 17$

13. $a = 46$, $b = 95$, $c = 7$, $d = 3.5$

Reteaching with Additional Practice

1. reflection in the line $x = -1$

2. $E(0, 0)$, $F(0, 2)$, $G(2, 3)$, $H(2, 0)$

3. *A* and *H*, *B* and *G*, *C* and *F*, *D* and *E*

4. no   5. yes   6. $x = 12$, $y = 4$

7. $x = 40$, $y = 4$

Real-Life Application

1. B to C: 90° clockwise; D to F: 90° counterclockwise; F to H: 90° counterclockwise; H to J:
90° counterclockwise; E to I: 90° counterclockwise; R to W: 180°

2. D and F: vertical line; D and J: horizontal
line; E and I: vertical or horizontal line; F and H:
horizontal line; H and J; vertical line; M and U:
horizontal line; N and Z: horizontal line; P and V:
horizontal line; R and W: vertical line; D and H:
diagonal line; F and J; diagonal line.

Challenge: Skills and Applications

1. *Sample answer:*

Statements	Reasons
1. Q is between P and R.	1. Given
2. $PQ + QR = PR$	2. Segment Addition Postulate
3. $PQ = P'Q'$, $QR = Q'R'$, $PR = P'R'$	3. Given
4. $P'Q' + Q'R' = P'R'$	4. Substitution prop. of equality
5. Q' is between P' and R'.	5. Segment Addition Postulate

2. *Sample Answer:* If $\angle PQR$ is a straight angle,
the desired result follows immediately from the
previous exercise. Therefore, we may assume that
P, Q, and R are noncollinear, so $\triangle PQR$ exists.
Since $PQ = P'Q'$, $QR = Q'R'$, $PR = P'R'$,
$\triangle PQR \cong \triangle P'Q'R'$ by the SSS Congruence
Postulate. Since corresponding parts of congruent
triangles are congruent, $\angle PQR \cong \angle P'Q'R'$.

3. $(3, -1)$ 4. $(6, 21)$ 5. $(8, -4)$ 6. $(-1, 9)$

7. $A'(5, 1)$, $B'(8, 0)$, $C'(6, -4)$; translation;
isometry   8. $A'(2, 8)$, $B'(8, 6)$, $C'(4, -2)$; other
transformation; not an isometry   9. $A'(1, -4)$,
$B'(4, -3)$, $C'(2, 1)$; reflection; isometry

10. $A'(4, -1)$, $B'(3, -4)$, $C'(-1, -2)$; rotation;
isometry   11. $A'(4, 4)$, $B'(1, 3)$, $C'(3, -1)$;
translation; isometry   12. $A'(5, -4)$, $B'(8, -3)$,
$C'(6, 1)$; translation; isometry

Lesson 7.2

Warm-Up Exercises

1. IV   2. II   3. Q lies on m and $\overline{PQ} \perp m$

4. SAS

Daily Homework Quiz

1. *FEHG*   2. reflection in the *y*-axis

3. $(4, 4)$ 4. No; the image is not congruent to
the preimage.   5. $w = 23\frac{1}{3}$, $x = 4$, $y = 7$

Lesson Opener

Allow 10 minutes.

1. The image extends a right arm up.

2. The image places its closer hand at the same
point "on" the mirror.

3. *Sample answer:*

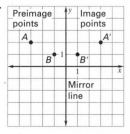

Lesson 7.2 *continued*

Technology Activity

1. Check student's work. **2.** Check student's work. **3.** Check student's work.

Practice A

1. not a reflection **2.** yes **3.** not a reflection
4. $\overline{CB}$ **5.** $\overline{CG}$ **6.** $\angle G$ **7.** $\angle BED$
8. $\angle FBC$ **9.** B **10.** $CGFB$

11. **12.**

13. 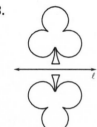 **14.** $(4, -2)$ **15.** $(3, -5)$
16. $(-5, 1)$ **17.** $(-3, 0)$

18–21. Answers vary; *Sample answers:*

18.   **19.**

20.   **21.**

Practice B

1. 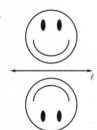 **2.**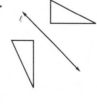

3.   **4.** false **5.** false
6. false **7.** true
8. $\triangle 4$ **9.** $\triangle 2$ **10.** $\triangle 3$
11. $\triangle 1$ **12.** $\triangle 3$

13–16. Answers vary; *Sample answers:*
13. not possible
14.   **15.** not possible

16.   **17.**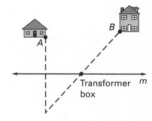

18. 4.0 cm

Practice C

1.   **2.**

3. 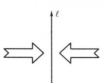 **4.** $(3, -2)$ **5.** $(-2, -4)$
6. $(-4, 3)$ **7.** $(1, -2)$
8. segment 7
9. segment 2

10. segment 4 **11.** segment 1 **12.** segment 4
13. segment 4 **14.** CHECKBOOK **15.** B, C, D, E, H, I, K, O, X **16.** Answers vary.
17. $(3, 0)$ **18.** $(2, 0)$ **19.** $(0, 0)$ **20.** $(-1, 0)$
21. $(-1, -1)$ **22.** 2 **23.** $-\frac{1}{2}$
24. $y = -\frac{1}{2}x - \frac{3}{2}$

Lesson 7.2 *continued*

Reteaching with Additional Practice

1.

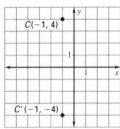

$C(-1, 4)$
$C'(-1, -4)$

2.

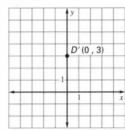

$D'(0, 3)$

3.

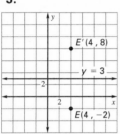

$E'(4, 8)$
$y = 3$
$E(4, -2)$

4.

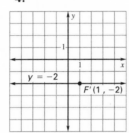

$y = -2$
$F'(1, -2)$

5.

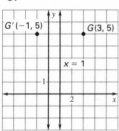

$G'(-1, 5)$ $G(3, 5)$
$x = 1$

6.

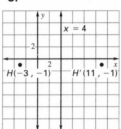

$x = 4$
$H(-3, -1)$ $H'(11, -1)$

7.

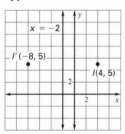

$x = -2$
$I'(-8, 5)$
$I(4, 5)$

8.

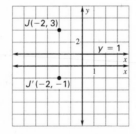

$J(-2, 3)$
$y = 1$
$J'(-2, -1)$

9. two **10.** one **11.** $(2, 0)$ **12.** $(5, 0)$
13. $(1.5, 0)$

Interdisciplinary Application

1. ; center line **2.** A cut made into the folded edge of the paper will create a design with four lines of symmetry.

3. Answers may vary. **4.** 8 lines of symmetry
5. $22.5°$

Challenge: Skills and Applications

1. *Sample answer:* The procedure is almost the same as the procedure for drawing a line that is perpendicular to k and passes through P. Use the compass with the point at P to sketch an arc that intersects k twice. Then sketch additional arcs of the same radius, with the compass point at the two previous intersection points. P' is the point opposite P, where the new arcs intersect.

2.

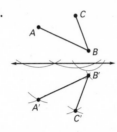

3.

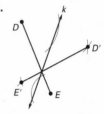

4. *Sample answer:* Since Y is the reflections of X in line m, m is the perpendicular bisector of $\overline{YX}$. By the Perpendicular Bisector Theorem, $OX = OY$ and $OY = OZ$. By the transitive property of equality, $OX = OZ$. **5.** $m\angle XOZ = 2m\angle SOT$; *Sample answer:* Since Y is the reflection of X in line m, m is the perpendicular bisector of $\overline{XY}$. Therefore, $\overline{SX} \cong \overline{SY}$ and, since $\angle OSX$ and $\angle OSY$ are right angles, $\angle OSX \cong \angle OSY$. Also, $\overline{OS} \cong \overline{OS}$ (Reflexive Property of Congruence) so $\triangle OSX \cong \triangle OSY$. Since corresponding parts of congruent triangles are congruent, $\angle XOS \cong \angle YOS$. By a similar argument, since Z is the reflection of Y in n, $\angle YOT \cong \angle ZOT$. By the Angle Addition Postulate and the substitution property of equality,

$m\angle XOZ$
$= m\angle XOS + m\angle YOS + m\angle YOT + m\angle ZOT$
$= m\angle YOS + m\angle YOS + m\angle YOT + m\angle YOT$
$= 2(m\angle YOS + m\angle YOT)$
$= 2m\angle SOT$. **6.** 3 **7.** 4 **8.** 5 **9.** 6 **10.** 2
11. 2 **12.** 1 **13.** 0 **14.** 1 **15.** 0

Lesson 7.3

Warm-Up Exercises

1. SAS **2.** subtraction property of equality
3. $180°$ **4.** $270°$

Daily Homework Quiz

1. $A'(3, 0)$ **2.** $B'(2, 4)$ **3.** Check drawings.
4. $(3, 0)$

Lesson 7.3 *continued*

Lesson Opener

Allow 15 minutes.

1. 4 stamps   **2.** Sample answers are given. See Exercise 1 for a pattern with a 90° angle of rotation.

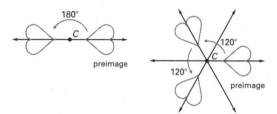

preimage

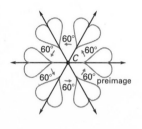

preimage

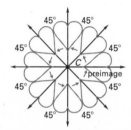

preimage

Angle of rotation	Number of stamps in pattern
180°	2
120°	3
90°	4
60°	6
45°	8

The product is always 360°.

Practice A

1. yes; a rotation of 60°, 120°, or 180° clockwise or counterclockwise about its center   **2.** yes; a rotation of 180° clockwise or counterclockwise about its center   **3.** yes; a rotation of 180° clockwise or counterclockwise about its center   **4.** no

5. B   **6.** E   **7.** B   **8.** F   **9.** F   **10.** D

11.

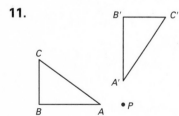

12.

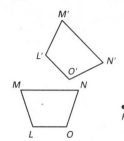

13.

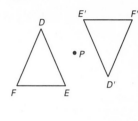

14. $A'(1, 3)$, $B'(1, 1)$, $C'(5, 1)$
15. $A'(-1, -1)$, $B'(-1, -3)$, $C'(-4, -3)$, $D'(-4, -1)$ **16.** $A'(-1, 0)$, $B'(-4, 0)$, $C'(-3, -3)$, $D'(-3, -2)$

Practice B

1. yes; a rotation of 90° or 180° clockwise or counterclockwise about its center   **2.** yes; a rotation of 60°, 120°, or 180° clockwise or counterclockwise about its center   **3.** yes; a rotation of 45°, 90°, 135°, or 180° clockwise or counterclockwise about its center   **4.** yes; a rotation of 45°, 90°, 135°, or 180° clockwise or counterclockwise about its center   **5.** $\overline{CD}$

6. $\overline{FG}$ **7.** $\overline{FE}$ **8.** $\overline{AB}$ **9.** $\triangle HPA$

10. $\triangle GPH$ **11.** 64°   **12.** 64°   **13.** 180°

14. $A'(1, 3)$, $B'(3, 3)$, $C'(3, -1)$, $D'(1, -1)$

15. $A'(2, 0)$, $B'(3, -2)$, $C'(2, -4)$, $D'(1, -2)$

16. $A'(-2, 0)$, $B'(-4, 1)$, $C'(-4, 5)$, $D'(-2, 4)$

Practice C

1. yes; a rotation of 90° or 180° clockwise or counterclockwise about its center   **2.** yes; a rotation of 45°, 90°, 135°, or 180° clockwise or counterclockwise about its center   **3.** yes; a rotation of 45°, 90°, 135°, or 180° clockwise or counterclockwise about its center

4. yes; a rotation of 72° or 144° clockwise or counterclockwise about its center   **5.** $\overline{CD}$

6. $\overline{FG}$ **7.** $\overline{EF}$ **8.** $\overline{AB}$ **9.** $\triangle GLH$

10. $\triangle EKD$ **11.** 76°   **12.** 106°

13. $A'(1, 3)$, $B'(1, 5)$, $C'(4, 6)$, $D'(4, 2)$

14. $A'(2, -1)$, $B'(5, -1)$, $C'(4, -3)$, $D'(1, -3)$

15. $A'(-4, 0)$, $B'(-6, 2)$, $C'(-4, 6)$, $D'(-2, 2)$

Answers

Lesson 7.3 *continued*

Reteaching with Additional Practice

1. $A'(2, 2), B'(-1, -2), C'(5, -2), D'(0, -6)$
2. $X'(-2, -5), Y'(-5, -5), Z'(-5, -1),$ $W'(-2, -1)$ 3. Yes, a rotation of 120° about its center   4. Yes, a rotation of 180° about its center
5. Yes, a rotation of 72° about its center

Real-Life Application

1. 40°   2. 4   3. 5   4. 5 minutes

Math and History Application

1. 4   2. Yes; the floor map can be rotated 90° in either direction.

Challenge: Skills and Applications

1. *Sample answer:* Draw rays $\overrightarrow{PX}$ and $\overrightarrow{PX'}$. Draw a circle, centered at P, of radius PS. Set the compass radius using the intersections of this circle with the rays you drew. Then, with the compass point at S, you can draw an appropriate arc whose intersection with the circle is S'.

2.
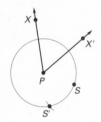

3.
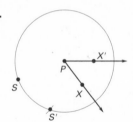

4. $(x_0, y_0 + a - x_0)$
5. $(x_0 - b + y_0, y_0 + a - x_0)$ 6. $(-3, 5)$
7. $(9, 12)$ 8. yes; 120°   9. yes; 90°, 180°
10. yes; 72°, 144°   11. yes; 60°, 120°, 180°
12. yes; 180°   13. yes; 180°   14. no   15. no

Quiz 1

1. $XYZW$ 2. reflection in a line   3. Yes; all lengths, angle measures, parallel lines and distances are preserved.   4. $(1, -3)$ 5. $(2, -3)$
6. $(-2, 0)$ 7. $(-5.2, -2)$ 8. The arc $\overset{\frown}{AB}$ is rotated about the x-axis at angles of 90°, 180°, and 270°. The arc $\overset{\frown}{AB}$ is also reflected in the y-axis and the new arc $\overset{\frown}{A'B'}$ is rotated also at angles of 90°, 180°, and 270°.

Lesson 7.4

Warm-Up Exercises

1. $K'(1, 3), L'(3, 4), M'(-2, 6)$
2. $K'(3, -2), L'(5, -1), M'(0, 1)$
3. 5 units   4. 5 units

Daily Homework Quiz

1. Check drawings.
2. $A'(-2, -2), B'(1, -4), C'(1, -5)$
3. 50°

Lesson Opener

Allow 10 minutes.

1. Elevator: Figure 2; Escalator: Figure 3; Moving walkway: Figure 1

2. *Sample answer:* The elevator moved Abigail vertically (down 3 floors). The escalator moved Abigail both horizontally (an unknown distance) and vertically (up 1 floor), along a straight path that slants upward. The moving walkway moved Abigail horizontally (past 16 gates).

Practice A

1. translation   2. translation   3. rotation
4. translation   5. C   6. A   7. B   8. $(5, -5)$
9. $(1, -2)$ 10. $(9, -4)$ 11. $(-4, 10)$
12. $(7, -6)$ 13. $(-2, 2)$

14.
$A'(-1, 7), B'(3, 7),$
$C'(4, 3), D'(0, 3)$

15.
$A'(-4, 1), B'(0, 1),$
$C'(1, -3), D'(-3, -3)$

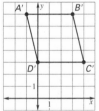

16.
$A'(-7, 9), B'(-3, 9),$
$C'(-2, 5), D'(-6, 5)$

17.
$A'(1, -2), B'(5, -2),$
$C'(6, -6), D'(2, -6)$

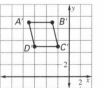

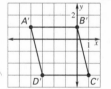

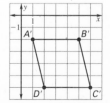

Lesson 7.4 *continued*

Practice B

1. a. $(x, y) \rightarrow (x + 5, y + 2)$ **b.** $\langle 5, 2 \rangle$

2. a. $(x, y) \rightarrow (x - 7, y - 2)$ **b.** $\langle -7, -2 \rangle$

3. $\overrightarrow{MN}, \langle 4, -2 \rangle$ **4.** $\overrightarrow{PQ}, \langle 4, 1 \rangle$

5. $\overrightarrow{RS}, \langle 0, -4 \rangle$ **6.** $(-1, 10)$ **7.** $(-6, 13)$

8. $(2, -12)$ **9.** $(12, -13)$ **10.** $(-5, 10)$

11. $(1, -2)$

12. $A'(-4, 3),$ **13.** $A'(1, -1),$
$B'(-1, 4), C'(1, 0)$ $B'(4, 0), C'(6, -4)$

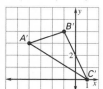
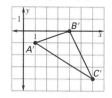

14. $A'(3, 1),$ **15.** $A'(-6, 0),$
$B'(6, 2), C'(8, -2)$ $B'(-3, 1), C'(-1, -3)$

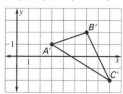
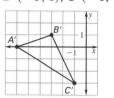
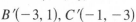

16. 8 **17.** 8

Practice C

1. $(x, y) \rightarrow (x + 4, y + 3)$

2. $(x, y) \rightarrow (x - 5, y - 2)$

3. $(x, y) \rightarrow (x - 1, y + 1)$

4. $(x, y) \rightarrow (x, y - 3)$

5. $(x, y) \rightarrow (x - 7, y - 4)$

6. $(x, y) \rightarrow (x + 10, y + 8)$

7. $\overrightarrow{MT}, \langle 4, 2 \rangle$ **8.** $\overrightarrow{JD}, \langle 5, -1 \rangle$

9. $\overrightarrow{LW}, \langle -4, -2 \rangle$

10. $\langle 5, -6 \rangle; B'(11, -4), C'(8, -8)$

11. $\langle 0, 6 \rangle; A'(-2, 10), B'(6, 8)$

12. $\langle -3, 1 \rangle; B'(3, 3), C'(0, -1)$

13. $\langle -4, -7 \rangle; A'(-6, -3), C'(-1, -9)$

14. $\langle -7, -3 \rangle; A'(-9, 1), B'(-1, -1)$

15. $\langle 2, 4 \rangle; A'(0, 8), C'(5, 2)$

16.
$M'(-3, 7), N'(-1, 3),$
$O'(2, 2), P'(5, 6)$

17.
$M'(4, 4), N'(6, 0),$
$O'(9, -1), P'(12, 3)$

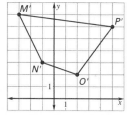
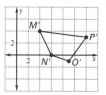

18. $M'(-4, -4), N'(-2, -8),$
$O'(1, -9), P'(4, -5)$

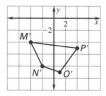

19. $M'(3, -2), N'(5, -6),$
$O'(8, -7), P'(11, -3)$

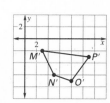

20. By the definition of a reflection, we know that
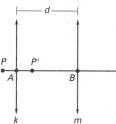
k is the $\perp$ bisector of $\overline{PP'}$. Let A be the point where $\overline{PP'}$ intersects k. By definition of $\perp$ bisector, $PA = P'A$. Similarly, if B is the point where $P'P''$ intersects m, $P'B = P''B$. By Segment Addition Postulate, $PP'' = PP' + P'P'', PP' = PA + P'A$ and $P'P'' = P'B + P''B$. Now substitute to show $PP'' = PA + P'A + P'B + P''B$. Substitute again and simplify to show $PP'' = 2P'A + 2P'B$ and $PP'' = 2(P'A + P'B)$. We are given that d is the distance between lines k and m. Since $d = AB$ and by Segment Addition Postulate $AB = P'A + P'B$, substitute to show $PP'' = 2(AB)$ or $PP'' = 2d$.

Reteaching with Additional Practice

1. XX', ZZ' **2.** 3 **3.** $X''Y''Z''$ **4.** k and m

Answers

Lesson 7.4 *continued*

5.

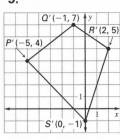

Q'(−1, 7)
R'(2, 5)
P'(−5, 4)
S'(0, −1)

6.

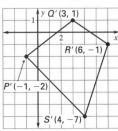

Q'(3, 1)
R'(6, −1)
P'(−1, −2)
S'(4, −7)

7.

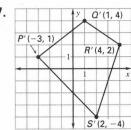

Q'(1, 4)
P'(−3, 1)
R'(4, 2)
S'(2, −4)

8.

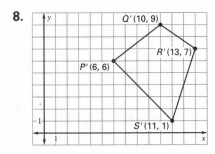

Q'(10, 9)
R'(13, 7)
P'(6, 6)
S'(11, 1)

9. $(x − 2, y − 4)$ **10.** $(x + 3, y − 3)$

Cooperative Learning Activity

1. Pulling perpendicular requires less force.

2. *Sample answer:* The component form of the vector that represents pulling the object in the perpendicular direction combines the horizontal and vertical components. Therefore, it is larger than the horizontal component. As a result, less force is required to move the object if you pull the object in the perpendicular direction than if you pull the object in the parallel direction.   **3.** yes

Interdisciplinary Application

1. $2\sqrt{85}$ feet   **2.** $\langle 6, −7 \rangle, \langle 1, −5 \rangle$
3. $\langle 5, −2 \rangle$ **4.** $\langle 12, −14 \rangle$

Challenge: Skills and Applications

1. *Sample answer:* Since $XX'Y'Y$ is a parallelo-gram, $YY' = XX'$ and $X'Y' = XY$. So, with the compass point at Y, construct an arc of radius XX', and with the compass point at X', construct an arc of radius XY. Then Y' is an intersection of these two arcs.

2.

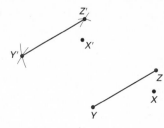

Z'
Y'
X'
Z
X
Y

3.

Y'   Z'
Y   Z   X'
• X

4. true; *Sample answer:* Let $P = (a, b)$ and $Q = (c, d)$. Then $P' = (a + h, b + k)$ and $Q' = (c + h, d + k)$. Then $PP' = QQ'$ $= \sqrt{h^2 + k^2}$, so $\overline{PP'}$ and $\overline{QQ'}$ are congruent.

Also, both $\overline{PP'}$ and $\overline{QQ'}$ have slope $\dfrac{k}{h}$ (or undefined slope, if $h = 0$), so they are parallel (or collinear).   **5.** false: *Sample answer:* Consider a rotation about the origin, where P is $(2, 0)$ and Q is $(−2, 0)$. Then $\overline{PP'}$ and $\overline{QQ'}$ are parallel and congruent, but the isometry is not a translation.

6. h is a multiple of 90; k is a multiple of 50.

Lesson 7.5

Warm-Up Exercises

1. $P'(−4, −1), Q'(−1, 0)$
2. $P'(−3, −5), Q'(0, −4)$ **3.** $P'(4, 4), Q'(1, 3)$
4. $P'(−4, 4), Q'(−1, 3)$

Daily Homework Quiz

1. a. $(x, y) \rightarrow (x − 4, y − 3)$ **b.** $\langle −4, −3 \rangle$
2. $\overrightarrow{AB} = \langle 3, 3 \rangle$ **3. a.** $(6, 4)$ **b.** $(−5, 4)$
4. $\langle 3, −4 \rangle$

Lesson Opener

Allow 15 minutes.

1–2. Check student's work.

Lesson 7.5 *continued*

3. *Sample description:* Notice that the original flower is contained in a diagonal line of flowers that goes upward as it goes to the right. The image is the next flower up and to the right along that diagonal. The image leans to the left whereas the original flower leaned to the right.

4. The final images are the same.

5. *Sample answer:* The transformations in both exercises consisted of a translation and a reflection. The figure that resulted was the same. The order of the transformations in Exercises 3 and 4 was reversed.

Technology Activity

1. no   **2.** yes   **3.** yes

Practice A

1. D   **2.** B   **3.** A   **4.** C

5. $A'(-1, 2), B'(-4, 3)$

6. $A'(-2, 1), B'(-3, 4)$ **7.** $A'(-3, 5), B'(0, 6)$

8. $A'(3, 2), B'(0, 3)$

9. $A'(-1, -2), B'(-4, -3)$

10. $A'(6, -4), B'(9, -3)$

11. $A'(3, 3)$

12. $A'(-3, 4)$

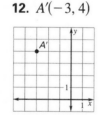

13. $A'(6, -1)$

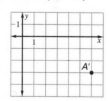

14. $A'(6, -3)$

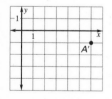

15. Reflection in x-axis followed by translation of $(x, y) \rightarrow (x + 6, y + 2)$.   **16.** Rotation of 90° counterclockwise about the origin followed by a reflection over the x-axis.

Practice B

1. C   **2.** A   **3.** B

4. $A'(-1, -1), B'(-1, -4), C'(2, -2)$

5. $A'(5, -1), B'(2, -1), C'(4, 2)$

6. $A'(4, -3), B'(7, -3), C'(5, 0)$

7. $A'(-1, 1), B'(-4, 1), C'(-2, -2)$

8. $A'(-2, -3), B'(1, -3), C'(-1, 0)$

9. $A'(-1, 1), B'(-1, 4), C'(2, 2)$

10. $A'(4, 5)$ **11.** $A'(-6, 1)$

12. $A'(4, 3)$ **13.** $A'(4, -1)$

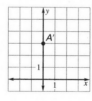

14. Reflection in $y = \frac{1}{2}$ followed by rotation about $(1, -3)$.   **15.** Rotation about $(-2, -1)$ followed by a translation $(x, y) \rightarrow (x + 5, y - 2)$.

Practice C

1. $A'(3, -1), B'(-1, 1), C'(1, 5)$

2. $A'(-3, -3), B'(-5, 1), C'(-9, -1)$

3. $A'(5, 1), B'(7, 5), C'(11, 3)$

4. $A'(3, -1), B'(-1, 1), C'(1, 5)$

5. $A'(-6, -7), B'(-4, -3), C'(0, -5)$

6. $A'(3, 1), B'(-1, -1), C'(1, -5)$

7. $A'(0, 3)$ **8.** $A'(17, 5)$

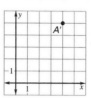

9. $A'(2, -5)$ **10.** $A'(-1, 4)$

11. Translation $(x, y) \rightarrow (x + 5, y - 4)$ followed by reflection in x-axis.   **12.** Rotation about $(0, 2)$ followed by translation $(x, y) \rightarrow (x, y + 4)$.

Lesson 7.5 *continued*

13. yes; $\langle 0.05, 0 \rangle$ **14.** no   **15.** yes; $\langle 0.025, 0 \rangle$
16. yes; $\langle 0.025, 0 \rangle$

Reteaching with Additional Practice

1. does not affect the image
2. does affect the image
3. does affect the image
4. does not affect the image

5.
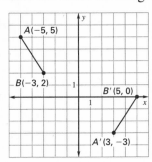

6.
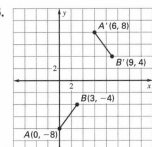

7.
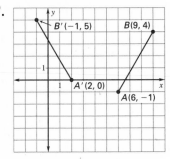

8.
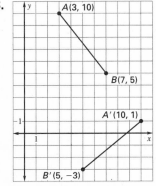

Real-Life Application

1. translation, reflection over a vertical line, and a 90° rotation counterclockwise   **2.** translation and a reflection over a horizontal line

3. translation and a 90° rotation counterclockwise   **4.** translation, a reflection over a vertical line, and a 90° rotation clockwise

Challenge: Skills and Applications

1. $(x, y) \rightarrow (x + 6, y - 4)$
2. $(x, y) \rightarrow (x + 9, y - 6)$
3. $(x, y) \rightarrow (x + 12, y - 8)$
4. $(x, y) \rightarrow (x + 3n, y - 2n)$
5. $(x, y) \rightarrow (x + 6, y - 4)$
6. $(x, y)(x + 6, y - 4)$ **7.** $(x, y) \rightarrow (x + 6, -y)$
8. $(x, y) \rightarrow (x + 6, -y)$ **9.** $(x, y) \rightarrow (-x, y - 4)$
10. $(x, y) \rightarrow (-x, -y)$
11. $(x, y) \rightarrow (-x - 3, y - 2)$
12. $(x, y) \rightarrow (-x + 3, y - 2)$
13. $(x, y) \rightarrow (y, -x)$ **14.** $(x, y) \rightarrow (-x, -y)$
15. $(x, y) \rightarrow (-y, x)$ **16.** $(x, y) \rightarrow (x, y)$
17. $(x, y) \rightarrow (y, x)$ **18.** $(x, y) \rightarrow (-y, -x)$
19. $(x, y) \rightarrow (-y, -x)$
20. $(x, y) \rightarrow (y - 2, -x - 3)$ **21.** *RV, VR, SW*
22. $T^2 = RS = SR, RV = VR, VW = U^2$, $VU = UW$ **23.** *Sample answer: j: x = 0, k: x = 3*   **24.** *Sample answer: j: y = 0, k: y = -4*   **25.** *Sample answer: j: y = 2x, k: y = 2x + 5*   **26.** *Sample answer: j: y = 0, k: x = 0*   **27.** *Sample answer: j: y = 0, k: y = x*

Lesson 7.6

Warm-Up Exercises

1. C   **2.** D   **3.** A   **4.** B

Daily Homework Quiz

1. $A''(1, -2)$ **2.** $A''(2, 3)$

3. $\triangle ABC$ is reflected in line $x = 1$ to produce $\triangle A'B'C'$. Then $\triangle A'B'C'$ is rotated 90° clockwise about the point $(1, -1)$ to produce $\triangle A''B''C''$.

Lesson 7.6 *continued*

Lesson Opener

Allow 20 minutes.

1. 6 stitches; 14 times. This headband shows horizontal translation, 180° rotation, reflection in a horizontal line, reflection in a vertical line, and horizontal glide reflection. **2.** Answers will vary.

Practice A

1. THG **2.** TRVG **3.** TV **4.** T
5. TRHVG **6.** TR **7.** TG **8.** translation
9. translation, horizontal line reflection, horizontal glide reflection **10.** translation, vertical line reflection, horizontal glide reflection, 180° rotation **11.** translation, vertical line reflection, horizontal glide reflection, 180° rotation

12–15. Answers vary. *Sample answers:*

12.

13.

14.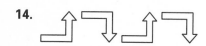

15.

16. TV or T **17.** TV **18.** TRVG **19.** TV

Practice B

1. yes; reflection in *x*-axis **2.** yes; reflection in any vertical line that lies midway between two figures **3.** yes; 180° rotation about (7, 0), (14, 0), (21, 0) and so on **4.** translation $(x, y) \rightarrow (x + 14, y)$ **5.** reflection over the *x*-axis **6.** rotation about (7, 0) **7.** glide reflection consisting of a reflection in the *x*-axis and a translation $(x, y) \rightarrow (x + 14, y)$

8–11. Answers vary; *Sample answers:*

8.

9.

10.

11. 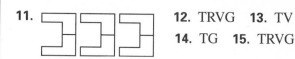 **12.** TRVG **13.** TV
14. TG **15.** TRVG

Practice C

1. yes; reflection in *x*-axis **2.** yes; reflection in any vertical line midway between two figures.

3. yes; 180° rotation about (6, 0), (12, 0), (18, 0), and so on **4.** translation $(x, y) \rightarrow (x + 12, y)$

5. reflection over the *x*-axis **6.** rotation about (6, 0) **7.** glide reflection consisting of a reflection in *x*-axis followed by a translation $(x, y) \rightarrow (x + 12, y)$

8–11. Answers vary; *Sample answers:*

8.

9.

10.

11.

12. T **13.** TV **14.** TRVG **15.** TRHVG

Reteaching with Additional Practice

1. TV or T **2.** T **3.** TV or T **4.** Yes, in the line $y = -\frac{1}{2}$ **5.** Yes, in the line $x = 6$, $x = 13$
6. reflection in the lines $x = 13$
7. horizontal glide reflection

Interdisciplinary Application

1. TR **2.** T **3.** TG, TV, THG, TRHVG; to meet the requirements for these frieze patterns, the notes would violate our rule of musical notation.

Challenge: Skills and Applications

1. translation; 180° rotation; horizontal and vertical glide reflection **2.** translation only
3. translation; reflection in vertical line; vertical glide reflection **4.** translation; 90° and 180° rotation **5.** translation; 90° and 180° rotation; reflection in vertical, horizontal, and diagonal

Lesson 7.6 *continued*

(45° from horizontal) lines; glide reflection along vertical, horizontal, and diagonal (45° from horizontal) lines   **6.** translation; 60°, 120°, and 180° rotation; reflection in vertical, horizontal, and diagonal (30° from either vertical or horizontal) lines; glide reflection along vertical, horizontal, and diagonal (30° from either vertical or horizontal) lines

Quiz 2

1. $A'(0, 3), B'(4, 7), C'(3, 2)$ **2.** $A'(-3, 4),$
$B'(1, 8), C'(0, 3)$ **3.** $A'(-2, -1), B'(2, 3),$
$C'(1, -2)$ **4.** $A'(3, 4), B'(7, 8), C'(6, 3)$

5.

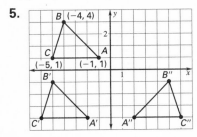

6.

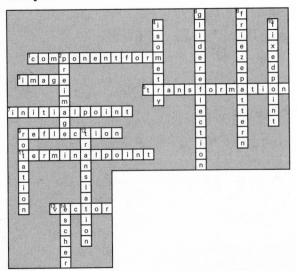

7. yes; TRHVG (Translation, 180° rotation, horizontal line reflection, vertical line reflection, horizontal glide reflection)

Review and Assessment

Chapter Review Games and Activities

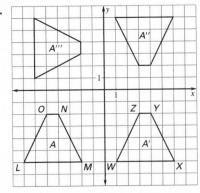

Test A

1. rotation   **2.** reflection   **3.** rotation
4. translation   **5.** 1   **6.** 1   **7.** 0   **8.** reflection
9. D **10.** S **11.** Answers may vary.
12. Answers may vary.   **13.** $\overrightarrow{AB}, \langle 4, 0 \rangle$
14. $\overrightarrow{CD}, \langle 4, 3 \rangle$ **15.** C   **16.** A   **17.** D   **18.** B
19. horizontal translation   **20.** horizontal translation and 180° rotation

Test B

1. rotation   **2.** reflection   **3.** rotation
4. translation   **5.** 2   **6.** 1   **7.** 0   **8.** reflection
9. D **10.** S **11.** Answers may vary.
12. Answers may vary.   **13.** $\overrightarrow{AB}, \langle 1, 3 \rangle$
14. $\overrightarrow{CD}, \langle 3, -3 \rangle$ **15.** C   **16.** A   **17.** D
18. B   **19.** horizontal translation, 180° rotation
20. horizontal translation, and vertical line reflection

Test C

1. rotation   **2.** translation   **3.** reflection
4. rotation   **5.** 1   **6.** 0   **7.** 2   **8.** reflection
9. I **10.** M **11.** Answers may vary.
12. Answers may vary.   **13.** $\overrightarrow{AB}, \langle 4, 3 \rangle$
14. $\overrightarrow{CD} = \langle -3, -5 \rangle$ **15.** D   **16.** A   **17.** C
18. B   **19.** translation and vertical line reflection   **20.** translation, horizontal line reflection, and glide reflection

SAT/ACT Chapter Test

1. B   **2.** A   **3.** C   **4.** D   **5.** C   **6.** B   **7.** E

Alternative Assessment

2. c. and e.

Review and Assessment *continued*

3. b.

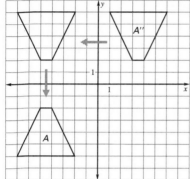

c.

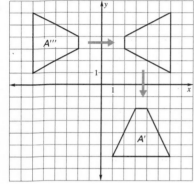

Project: Designing a Quilt Sample

1.–2. Check drawings.

3. Some students may draw diagonals or symmetry lines across the square and some may draw them part way across, stopping when they meet the first line. The directions allow for both possibilities. Many choices are possible for drawing the segments.   **4.** Check drawings to make sure the design does not exceed eight sections.

Cumulative Review

1. *B* **2.** *C* **3.** 30° **4.** 120° **5.** 11 **6.** 8

7.

Statements	Reasons
1. $\overline{AB} \parallel \overline{XY}$	**1.** Given
2. $\angle ABO \cong \angle XYO$	**2.** Alternate Int. $\angle$s Thm.
3. *O* is midpoint of $\overline{BY}$.	**3.** Given
4. $\overline{BO} \cong \overline{YO}$	**4.** Def. of midpoint
5. $\angle AOB \cong \angle XOY$	**5.** Vertical $\angle$s Thm.
6. $\triangle ABO \cong \triangle XYO$	**6.** ASA Congruence Postulate

8.

Statements	Reasons
1. $\angle A \cong \angle D$ $\angle B \cong \angle C$	**1.** Given
2. $\overline{BC} \cong \overline{BC}$	**2.** Reflexive prop. of congruence
3. $\triangle ABC \cong \triangle DCB$	**3.** AAS Congruence Postulate
4. $\overline{AB} \cong \overline{DC}$	**4.** Corresponding parts of $\cong \triangle$'s are $\cong$.

9. 4 **10.** 5 **11.** $\angle O$ **12.** $\overline{NO}$

13. $\overline{PM}$ **14.** true **15.** false **16.** *N*

17. *K* **18.** *M* **19.** $A(-6, 2), B(-2, 4), C(-1, 0)$ **20.** $A(-1, -1), B(3, 1), C(4, -3)$

21. $A(-2, 3), B(2, 5), C(3, 1)$

22. $A''(-3, -2), B''(-6, -6)$ **23.** $A''(-6, 2), B''(-1, -3)$

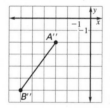 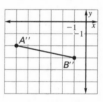

24.

25.